P9-BJM-612

FEARLESS SYMMETRY

FEARLESS

FEARLESS

Exposing the Hidden

Avner Ash *and*

PRINCETON UNIVERSITY PRESS

SYMMETRY

Patterns of Numbers

Robert Gross

PRINCETON AND OXFORD

Published by Princeton University Press, 41 William Street,
Princeton, New Jersey 08540
In the United Kingdom: Princeton University Press, 3 Market Place, Woodstock,
Oxfordshire OX20 1SY

Library of Congress Cataloging-in-Publication Data
Ash, Avner, 1949-
Fearless symmetry: exposing the hidden patterns of numbers /
Avner Ash and Robert Gross. p. cm.
Includes bibliographical references and index.
ISBN-13: 978-0-691-12492-6 (acid-free paper)
ISBN-10: 0-691-12492-2 (acid-free paper)
1. Number theory. I. Gross, Robert, 1959- II. Title.
QA241.A84 2006
512.7—dc22 2005051471

British Library Cataloging-in-Publication Data is available

About the jacket image, courtesy of Bahman Kalantari: Polynomiography is the art and science of visualization in approximation of the zeros of polynomial equations using iteration functions. Although its theoretical foundation can be traced to the well-known Fundamental Theorem of Algebra, polynomiography offers a new and exciting view into the world of polynomials as well as the mysteries of this beautiful theorem itself. Not only is polynomiography interesting and useful from the scientific and educational points of view, but it turns the ancient root-finding problem into a serious medium for creating artwork of great variety and diversity through a combination of human creativity and computer power. Each polynomial gives rise to an infinite number of 2D images, each called a *polynomiograph*. Each natural number can be identified as a polynomial. Hence, for each polynomial and each natural number there is an infinite number of polynomiographs waiting to be discovered. The particular image on the jacket is based on a polynomiograph produced by one of the techniques in polynomiography, referred to as *Voronoi coloring*. The title of the image is *Acrobats*. For more information, visit www.polynomiography.com.

This book has been composed in New Century Schoolbook

Printed on acid-free paper. ∞

pup.princeton.edu

Printed in the United States of America

1 3 5 7 9 10 8 6 4 2

FOR OUR PARENTS

FOR OUR PARENTS

Tyger, Tyger, burning bright,
In the forests of the night,
What immortal hand or eye
Could frame thy fearful symmetry?

—William Blake

In seed time learn, in harvest teach, in winter enjoy.

—William Blake, "Proverbs of Hell"

Contents

Foreword

At some point in his or her life every working mathematician has to explain to someone, usually a relative, that mathematics is hardly a finished project. Mathematicians know, of course, that it is far too soon to put the glorious achievements of their trade into a big museum and just become happy curators. In many respects, the study of mathematics has hardly begun. But, at least in the past, this has not always been universally acknowledged.

Recent successes (most prominently the proof of Fermat's Last Theorem) have advertised to a wide audience that math remains humanity's grand "work-in-progress," where mysteries abound and profound discoveries are yet to be made. Along with this has come a demand from a larger public for genuinely expository, but serious, accounts of currently exciting themes in mathematics.

It is a hard balancing act: to explain important and beautiful mathematical ideas—to *truly* explain them—to people with a general cultural background but no technical training in math, and yet not to slip away from the full seriousness and ambitious goals of the subject being explained.

Avner Ash and Robert Gross do a wonderful job with this balancing act in *Fearless Symmetry*. On the one hand the substance of their book is honestly—fearlessly, even—faithful to the great underlying ideas of the mathematical story that they tell. On the other hand, the authors are keenly sensitive to the basic, almost premathematical, issues that would occur to, and perhaps challenge, a newcomer to these ideas, and they treat these issues with an exemplary level of thoughtfulness.

The authors also bring out the *eternally unfinished* aspect of math, its open-ended quality. The resolution of any part of mathematics invariably modulates the subject into a different key, and makes a new and deeper set of questions vital. One theorem having been proved, more further-reaching problems come to prominence. Fermat's Last Theorem, posed over 350 years ago, has been proved; the curious Problem of Catalan, conceived over a century ago to prove that 8 and 9 are the only two consecutive perfect powers ($8 = 2^3; 9 = 3^2$), has recently been solved. But you need only glance at the last chapters of this book to see how, in the wake of the resolution of older problems, a new, and possibly richer, repertoire of interesting problems has come to occupy center stage, which would have astounded ancient Diophantus.

And waiting for future generations are the sweeping expectations posed by celebrated problems such as the ABC conjecture and the Riemann Hypothesis.

Fearless Symmetry begins where few math books do, with an enlightening discussion of what it means for one "thing" to *represent* another "thing." This action—deeming A a "representation" of B—underlies much mathematics; for example, counting, as when we say that these two mathematical units "represent" those two cows. What an extraordinary concept representation is and always has been. In Leibniz's essay *On the Universal Science: Characteristic* where he sketched his scheme for a universal language that would reduce ideas "to a kind of alphabet of human thought," Leibniz claimed his characters (i.e., the ciphers in his universal language) to be manipulable representations of ideas.

> All that follows rationally from what is given could be found by a *kind of calculus*, just as arithmetical or geometrical problems are solved.

Nowadays, whole subjects of mathematics are seen as represented in other subjects, the "represented" subject thereby becoming a powerful tool for the study of the "representing" subject, and vice versa.

The mathematics of *symmetry* also has had an astounding history. It timidly makes an appearance in Euclid's *Elements* under the guise of the notion of similarity (*to homoion*). In somewhat homey terms, the more modern attitude towards symmetry is that it is a geometric transformation that you can perform on an object that makes the object end up looking as if it were exactly the same, and in the same position, afterwards as before. For example, if we are working in Euclidean geometry, the symmetries of an equilateral triangle in the plane consist of the three flips about the angle bisectors through each of its vertices, and also the three (yes there are three!) rotations that preserve the figure (rotate around the center by 120 degrees, 240 degrees, and 0 degrees). By the end of the nineteenth century, with the emergence of Klein's "Erlangen program," the general notion of symmetry had established itself as the very foundations of geometry, since all homogenous geometric structure had come to be viewed as a consequence of the study of the groups of their symmetries.

Groups of symmetries of a geometric object possess an intrinsically algebraic structure, if we take the view that the *product* $S{\cdot}T$ of two symmetries S,T consists of the new symmetry gotten by first performing the symmetry T and then following that by performing the symmetry S. The surprise is that this kind of "multiplication structure" on the collection of symmetries of a geometry holds the key to a fuller understanding of that geometry. For example, from an understanding of the continuous family of all *congruences* of Euclidean geometry, together with knowledge of the corresponding multiplication as described above, we can (re)construct all of Euclid's geometry, with its straight lines, its angles, and its circles!

But after we have wrested these purely algebraic structures, *groups of symmetries*, from their geometric origins, we are entitled to consider them entirely as creatures in algebra, where they are called simply *groups*. We can also go the other way: to seek to *re-present* such an algebraic structure, a group, as a group of symmetries of some geometry; and, even more revealing perhaps, as a group of symmetries of a geometry *different* from the one from which it initially arose.

Viewed from this perspective, the bare algebraic notion of a *group* establishes itself as an emissary, of a sort, between different geometries: the same group might account for the symmetries of two disparate geometries. Even more relevant to the substance of *Fearless Symmetry* is the great legacy of Evariste Galois in the nineteenth century—that these algebraic entities, groups, may bridge the even wider divide between the algebra of equations and geometry: *A group of symmetries of some system of algebraic equations may be represented as the group of symmetries of some geometry*. This development is the underpinning of much modern number theory.

The first two parts of this book are devoted to all these underlying algebraic ideas, including an introduction to the wonderful world of modular arithmetic opened up to us by the genius of Gauss, modular arithmetic being the beginnings of what we will call, below, "local number theory."

The third, and last, part of this book points the reader to the frontiers. *Reciprocity laws* play a big role in this story, for they form the backbone of what present-day number theorists call "global number theory." A *local problem* is one that concerns itself with issues regarding divisibility by a single prime number p, or by its powers. "Global problems," in contrast, constitute the basic hard questions we wish to answer about whole numbers. Reciprocity laws, when available, represent the extra glue, the further constraint, in a problem of global number theory that ties together all the corresponding problems in the various local number theories connected to each of the prime numbers $p = 2, 3, 5, 7, 11, \ldots$.

Ash and Gross end their book with some comments on Fermat's Last Theorem. The celebrated proof of this theorem depends on the realization that a solution to the equation

$$X^p + Y^p = Z^p$$

with p an odd prime number and X, Y, Z nonzero integers, leads us to be able to find a very distinctive finite group of symmetries of algebraic numbers, and to be able to represent this group also as a group of symmetries of a specific finite geometry, this representation having peculiar properties. The demonstration of the impossibility of a nonzero integer solution X, Y, Z follows from the proof that these peculiar representations cannot, in fact, exist.

Fearless Symmetry can be read, at one level, by a reader who may have no particular mathematical experience but is interested in the important concepts that frame the mathematical viewpoint (e.g., the concept of *representation* in chapter 1, of *modular arithmetic* in chapter 4). Readers with some background in basic mathematics who are happy to do a few calculations and to make a few numerical experiments, will also gain much as they accompany the authors further in their examination of some of the mathematical structures that play a role in this fine book.

Barry Mazur

Preface

Mathematical research flourished in the twentieth century both in quality and quantity, and shows no signs of abating in the twenty-first century. Yet many people still have the misconceptions that

- everything important in mathematics has already been discovered, and
- mathematics is of interest only for its applications to science and technology.

Nevertheless, the general audience for current ideas in pure mathematics is clearly growing, as evidenced by a spate of recent books. Based on the level of some of these books, it seems clear that a segment of this reading public also desires to go more deeply into the mathematics than was typical a generation ago.

This book is a popular exposition of cutting-edge research in one important area of mathematics, number theory. In it, we hope both to share the excitement and to help increase popular awareness of the intrinsic beauty of contemporary explorations in pure mathematics.

We have in mind a broad audience, centered principally on those who have studied calculus. Though calculus is not used in this book, the amount of mathematical maturity needed to follow everything in our book probably requires that level of mathematical experience. Professional mathematicians who are not expert in number theory but who want to learn something of its latest methods should also find something worth reading here.

On the other hand, we include as potential readers those who have only studied some algebra.[1] For that reason we have explained many topics, such as complex numbers and modular arithmetic, which will be known to more advanced readers. In the later chapters, where we discuss more abstruse topics, we sometimes pause to explain things to readers with a more limited background.

This book follows a path to one particular area of modern number theory: generalized reciprocity laws. It will take most of the book to explain this concept. But here is an extremely brief description of the territory:

> We want to solve polynomial equations. Unlike the quadratic equation $ax^2 + bx + c = 0$, most equations *do not* have formulas that give you the solutions. Generalized reciprocity laws are very complicated algorithms that enable you to get crucial information about some of these more complicated equations. In favorable circumstances, they can be used to prove deep statements about the solutions sets of algebraic equations.

By the way, an equation doesn't have to be too complicated to lack a formula and require—if possible—a reciprocity law. For example, $x^5 + ax + b = 0$ lacks a formula. This surprising fact was proved in the early 1800s by the Italian mathematician Paolo Ruffini and the Danish mathematician Niels Henrik Abel and led to the concept of the Galois group. One way to get mileage out of the Galois group is to represent it in terms of either permutations or matrices. These representations encapsulate the patterns referred to in the subtitle of our book. From the representations we go on to construct reciprocity laws. All of these terms—equation, group, Galois group, permutation, matrix, representation, reciprocity law, and many others—are defined and discussed in the course of this volume.

We can now explain the meaning of our title, although in a sense, the full explanation requires the whole book. Some number patterns, like even and odd numbers, lie on the surface. But the more

[1]The few bits that mention trigonometry, logarithms, infinite series, or differentiation can be skipped without impairing your ability to follow the rest of the book.

you learn about numbers, both experimentally and theoretically, the more you discover patterns that are not so obvious. For one example of this, see chapter 7 on quadratic reciprocity.

After a hidden pattern is exposed, it can be used to find more hidden patterns. At the end of a long chain of patterned reasoning, you can get to very difficult theorems, exploring facts about numbers that you otherwise would not know were true. An example of this is Fermat's Last Theorem, the proof of which is briefly sketched in chapter 22. Fermat's Last Theorem is a sort of "negative pattern," saying that you can never find two perfect nth powers that sum to a perfect nth power if n is greater than 2.

There are many kinds of hidden numerical patterns, and most of the ones we deal with in this book can be encoded in Galois groups. Galois groups describe symmetries among solutions to integral polynomial equations—hence the word "symmetry" in our title.

The striking beauty of mathematics, which seems to shine from beyond the purely human world, suggested to us the magnificence of the Tyger in William Blake's poem. Taking off from Blake's epithet of the Tyger as evincing a "fearful symmetry," we characterize the mathematical symmetries of this book as "fearless." They may baffle you, but they won't bite you. Moreover, the intellectual bravery of the mathematical pioneers, both of the past and of the present—of Carl Friedrich Gauss, Andrew Wiles, and many others—surely merits the description of "fearless," as most people who have taken a difficult math class would testify.

Our book is divided into three parts. Most of the first part is devoted to definitions and constructions in modern algebra and number theory on a basic level. We call attention to the first chapter, which discusses the mathematical idea of *representation* in a general way that does not occur in elementary courses. If the reader finds this chapter too dry or overly abstract, it can be skipped on a first reading; however, the definitions in chapter 1 should be looked at. Chapter 6 contains information about solving Diophantine equations,[2] which may be new to many readers.

---

[2]These are polynomial equations with integer coefficients when we look for solutions where the variables are only allowed to take rational values.

The second part of the book is devoted to more advanced topics from algebra and number theory. The first three chapters of this part are on three very different subjects: Galois theory, elliptic curves, and matrices. The ensuing chapters begin to put the various ideas together: groups of matrices and representations of groups.

We then discuss polynomials and their Galois groups, and how all these Galois groups glue together into a big group, called "the absolute Galois group of the rational numbers." We introduce representations of the Galois group and explain just what features of these representations will be useful in our construction of reciprocity laws. We begin to see how to construct interesting representations.

The third and last part is devoted to our main goal: reciprocity laws that equate the traces of Frobenius under representations of the Galois group with values computed from some other mathematical object. These terms cannot be understood until the reader has absorbed the first two parts. We give many examples of these reciprocity laws and their applications, culminating in the proof by Andrew Wiles of Fermat's Last Theorem. We finish the book with an attempt to look ahead to what might be just beyond the frontier in this kind of number theory.

Interspersed throughout these discussions are various digressions. In them, we step back and look at the mathematics we are doing from a larger perspective, to see how it fits in with other parts of mathematics, or with science, or with the world at large.

In most chapters, we will suggest optional exercises for the reader. Doing them is not necessary to follow the text at a general level. Some exercises have their solutions given. Those which do not should be straightforward.

You can skip around or read the book straight through. Up until chapter 12, the chapters are fairly independent of one another. If something is not clear, it may be explained in the next page or two. If that doesn't help, try rereading the section or chapter. If it remains unclear, please let us know and we will try to fix it for the next edition.

At the beginning of each chapter we have put a "road map" to tell you where we are in reference to where we've been and where we are going. We have set off certain definitions for ease of reference. This is not a textbook, and we have not attempted to make it logically self-contained. The definitions are there to clarify usage, fix ideas, and resolve possible ambiguities. Mathematical definitions are not like dictionary definitions. They do not *describe* usage, but *create* it. A mathematical definition cannot be "wrong" or "right," only "inconsistent" or "consistent," "useless" or "useful."

We have stated a number of theorems along the way, mostly without proof. There are a few theorems with proofs, and some of the proofs can be quite involved. You do not need to follow the proofs in order to enjoy the rest of the book.

Acknowledgments

This book began as an expository article for professional mathematicians (Ash and Gross, 2000) about the proof of Fermat's Last Theorem. We received so many comments about that article that we decided to expand it into a book, filling in the necessary background so that more readers would be able to understand the beauty of the mathematics involved.

Along the way, our book was enormously improved by the comments of many people, among them Fedor Andrianov, Alexandra Ash, Ellie Ash, Jonathan Reiss, Rosemary Reiss, Karl Rubin, and two anonymous readers. We thank Henri Darmon for help with the chapter on Fermat's Last Theorem, and Klaus Peters for suggesting the idea of "road maps" to keep the reader oriented, as well as for many other helpful criticisms. Many thanks to Barry Mazur for his Preface. We are grateful to David Rohrlich for answering various number theory questions of ours over the years. We give special thanks to our ever-helpful editors, Vickie Kearn and Terri O'Prey, and to our excellent copyeditor, Steven Pisano.

Finally, we would like to thank our immediate families for their loving support during the writing of this book. They have provided us with environments in which our work could flourish.

Greek Alphabet

Alpha	A	α	Iota	I	ι	Rho	P	ρ
Beta	B	β	Kappa	K	κ	Sigma	Σ	σ
Gamma	Γ	γ	Lambda	Λ	λ	Tau	T	τ
Delta	Δ	δ	Mu	M	μ	Upsilon	Υ	υ
Epsilon	E	ϵ	Nu	N	ν	Phi	Φ	ϕ
Zeta	Z	ζ	Xi	Ξ	ξ	Chi	X	χ
Eta	H	η	Omicron	O	o	Psi	Ψ	ψ
Theta	Θ	θ	Pi	Π	π	Omega	Ω	ω

PART ONE

Algebraic Preliminaries

Chapter 1

REPRESENTATIONS

Road Map

To start our journey, we discuss the basic concept of *representation* from a formal point of view. This is the key concept underlying the number-theoretic methods of *Galois representations* that are our goal. To flesh out the abstract formalism, we go through an example: The ordinary act of counting can be viewed as a representation of sets. So we give (or review) mathematical definitions of *sets*, *functions*, *morphisms*, and *representations*, which will be with us for the whole book.

The Bare Notion of Representation

Before we narrow our focus to mathematical concepts, we start by discussing the general concept of a representation. In philosophy, the concept of one thing representing or misrepresenting another thing is a central concern. The distinction between truth and appearance, the thing-in-itself and its representation, is a keynote of philosophy. It plays a critical role in the works of such figures as Plato, Kant, Schopenhauer, and Nietzsche. Generally speaking, for these philosophers the "appearance" of something is thought to be an impediment or veil, which we wish to penetrate through to the reality acting behind it. But in mathematics, matters stand somewhat differently.

Consider, in an abstract way, the relationship that occurs when one thing represents another. Say B represents A. We have three terms that stand together in some kind of relationship: A, B, and the fact that B represents A. We can call this fact X. It is important to remember that, in a representation, the three terms A, B, and X are usually distinct.

For example, A may be a citizen of Massachusetts, B her state representative, and X the legal fact that B represents A by voting in the legislature on her behalf. Or, to jump ahead, A may be an abstract group, B a group of matrices, and X a morphism from A to B. (We will define these terms later.)

It can happen, though, that $A = B$. For instance, B may be said to (also) represent herself in the state legislature. Or A may be a group of matrices and B the same group of matrices. But whether $A = B$ or $A \neq B$, we call these relationships "representations."[1] Note that the fact of representation, X, is always going to be different from A and B, because A and B are objects and X is a fact of representation.

Now, what would be a good picture of A, B, and X? We can view X as an arrow going from A to B. This captures the one-way quality of the relationship, showing that B is representing A, not vice versa:

$$A \longrightarrow B.$$[2]

We can abstract even further, if we do not want to name A and B and we just want to visualize their relationship. We can picture them with dots. Then the picture of a representation becomes

$$\bullet \longrightarrow \bullet$$

which is the ultimate in abstraction. The dots are just placeholders for the names of the objects. The two dots can stand for two different objects or the same object. The dot or object from which the arrow emanates is called the *source* of that arrow, and the dot or object to which the arrow goes is called the *target* of that arrow.

In normal life, if A represents B, B and A can be very different kinds of things. For instance, a flag can represent a country, a

---

[1]It may not seem to make sense for an object to represent itself, or it may seem like the best, most exact possible representation. Mathematicians do not take sides in this debate. We just agree to call it a representation even when A represents itself.

[2]It *could* happen that, at the same time, A also represents B, and we would picture that as $B \longrightarrow A$. But this is a different representation from the previous one. Its "fact of representation" Y is not equal to X.

slogan on a T-shirt can represent an idea, and a mental image can represent a beloved person. In mathematics, the situation is different. All the mathematical entities we encounter or invent are considered to be on the same plane and have the same degree and type of reality or ideality: They are all mathematical entities.

What are representations used for? They explain one thing by means of another. The object we want to understand is the "thing": the thing-in-itself, the source. The object that we know quite a bit about already, to which we compare the source via a representation, we call the *standard object*. It is the site of appearance, the target.

Our conventions might not correspond to your expectations. The target, the object at the head of the arrow, is the piece of the picture that we understand better. We will derive information about the source by using properties of both the arrow and the target.

An Example: Counting

We look at the simplest possible example, one that goes back to prehistory: counting. Suppose we have a sack of potatoes or a flock of sheep. We want to know how many potatoes or sheep we have.

This is a much more sophisticated question than knowing whether they are the same in number as another sack of potatoes or another flock of sheep. We start with the less sophisticated question. Suppose we want to know whether the flock of sheep being herded home this evening is the same size as the herd we let out to the pasture in the morning. In the morning, we put a small pebble in our pouch for each sheep as it went out of the fold. Now we take a pebble out of the pouch as each sheep returns to the fold.

We were careful to make sure the pouch was empty in the morning before we began, and careful not to put anything in or take anything out during the day. So if the pouch becomes empty exactly as the last sheep comes in, we are happy. A mathematician says that we have demonstrated the existence of a *one-to-one correspondence* from the sheep in the morning to the sheep in the evening.

To make this mathematically precise, we make two definitions:

DEFINITION: A *set* is a collection of things, which are called the *elements* of the set.[3]

For example, the collection of all odd numbers is a set, and the odd number 3 is an element of that set.

DEFINITION: A *one-to-one correspondence* from a set A to a set B is a rule[4] that associates to each element in A exactly one element in B, in such a way that each element in B gets used exactly once, and for exactly one element in A.

Digression: Definitions

A mathematician uses the term "definition" in a way that might be surprising to nonmathematicians. The *Oxford English Dictionary* defines "definition" as "a precise statement of the essential nature of a thing." Mathematicians agree that a definition should be "precise," but we are not so sure about capturing the "essential nature." Our definition of one-to-one correspondence above will let you recognize a one-to-one correspondence if one is shown to you. Suppose that A is the set {red, blue, green} and B is the set $\{1, 2, 3\}$. Then a one-to-one correspondence between the two sets is given by

$$\text{red} \to 1$$

$$\text{blue} \to 2$$

$$\text{green} \to 3.$$

You can check that this associates to each element of the set A a different element of the set B, and that each element of the set B is used once.

[3]A set may be described by listing all of its elements between curly braces, so that $\{1, a, b\}$ is the set with the three elements $1, a$, and b. A set may also be described using a qualifier preceded by a colon, so that $\{x : x > 0 \text{ and } x \text{ is real}\}$ is the set of all positive real numbers.

[4]By "rule," we mean any definite means of association. It need not be given by a formula. For example, it can be given by a list that tells which sheep in the morning and in the evening were counted by the same pebble.

Our definition of one-to-one correspondence, however, does not tell you the "essential nature" of a one-to-one correspondence. We have given you no clue why you should care about one-to-one correspondences, nor does our definition tell you how to make a one-to-one correspondence.

Even when a mathematical definition technically has all of the properties listed by the *OED*, it often strikes nonmathematicians as unusual. A mathematical definition can redefine a commonly used word to mean something else. For example, mathematicians refer to "simple" groups, which are in fact not particularly simple. They define the words "tree" and "quiver" in ways that have nothing to do with oaks and arrows.

Sometimes a mathematician defines an object in terms of its properties, and only then proves that an object with these properties exists. Here is an example: The *greatest common divisor* of two positive integers a and b can be defined to be a positive number d so that:

1. d divides a.
2. d divides b.
3. If c is any other number that divides both a and b, then c divides d.

With this definition, it is not obvious that the greatest common divisor exists, because there might not be any number d that satisfies all three properties. So right after making the definition, it should be proved that a number with the properties outlined actually exists.

Counting (*Continued*)

In our example, each pebble corresponded to one sheep in the morning and one sheep in the afternoon. This sets up the rule that associates to each morning sheep the afternoon sheep that shared its pebble. This rule is a one-to-one correspondence under the conditions of our story.

But we do not need to know any set theory, nor what a one-to-one correspondence is, to count sheep in this way. In fact, we do not even need to know how to count! In a book about Sicily in the 1950s (Dolci, 1959), a young shepherd boy was interviewed who did not know how to count:

> I can't count, but even when I was a long way away, I could see if one of my goats was missing. I knew every goat in my herd—it was a big herd, but I could tell every one of them apart. I could tell what kid belonged to what mother. . . . The master used to count them to see if they were all there, but I knew they were all there without counting them.

You can see that for the shepherd boy, counting was not necessary. Nor is it required if we want to sell our flock for one dollar a sheep. We just pair up the dollars and the sheep. And in the case of two sacks of potatoes, we can take one potato out of each sack and throw the pair of potatoes over our shoulders. We repeat until one sack is empty. If the other sack is also empty, we have confirmed that there were the same number of potatoes in each sack to begin with.

Counting Viewed as a Representation

But if there are thousands of potatoes, or if we want to keep a record, or tell someone far away how many sheep we have, something else needs to be done, involving language—in this case, mathematical language. The flock of sheep is our "thing," our *source object*. For a target, we need a standard object that we know how to count in a standard way. This is the series of counting words, for example, in English, "one, two, three, . . . " As each sheep enters the fold, we count it with the next word in the series, and the last counting word that we utter is the number of sheep.

Again we have made a one-to-one correspondence, but this time with a standard object, so we have something to write home about. The folks at home have the same standard, so they will know how to interpret our report. (If we report our result to people who do not know the English counting words, they will not know how many sheep we have.)

In the case of the two sacks of potatoes, if we use the tossing-over-our-shoulders method, when we are done we will know whether the sacks contained the same number of potatoes or not, but the place will be strewn with potatoes and we will not know what that number is. If instead we use counting words, we can count the potatoes one sack at a time, neatly, and then compare the answers.

The Definition of a Representation

A one-to-one correspondence is an example of a *function* and of a *morphism*. We will be using these terms throughout this book. We will take a stab at defining them now, and refine and amplify the definitions as we continue.

> **DEFINITION**: A *function* from a set A to a set B is a rule that assigns to each element in A an element of B. If f is the name of the function and a is an element of A, then we write $f(a)$ to mean the element of B that is assigned to a. A function f is often written as $f : A \to B$.

> **DEFINITION**: A *morphism* is a function from A to B that captures at least part of the essential nature of the set A in its image in B.

We must be intentionally vague in this chapter about the way that a morphism "captures the essential nature" of A, mostly because it depends on the nature of the entities A and B. When we use the word "morphism" later in the book, our source A and target B will both be groups. After we have defined "group" in chapter 2, we will revisit the idea of a "morphism of groups" in chapter 12.

Some people may think "morphism" is an ugly word, but it is the standard mathematical term for this concept. The longer word "homomorphism" is also used, but we will stick with the shorter version. It derives from the Greek word for "form," and we view the "essential nature" captured by a morphism as the "form" of A.

There are many kinds of functions, but the most useful ones for us are the morphisms from a source to a well-understood standard target. We will call this a *representation*. It is implicit that the target we choose is one that we know a lot about, so that from our knowledge that there is a morphism, and better yet our knowledge of some additional properties of the morphism, we can obtain new knowledge about the source object.

DEFINITION: A *representation* is a morphism from a source object to a standard target object.

Counting and Inequalities as Representations

Going back to the counting example, we think about finite sets—for example, {sun, earth, moon, Jupiter} or $\{1, \text{Kremlin}, \pi\}$ or any set that contains a finite number of items. This collection of finite sets contains the special sets $\{1\}$, $\{1,2\}$, $\{1,2,3\}$, and so on. In the context of counting, given any two finite sets A and B, a morphism is a one-to-one correspondence from A to B. A representation in this case is a morphism from the source (a given finite set, e.g., the set of sheep in your flock) to the target, which must be one of the special sets $\{1\}$, $\{1,2\}$, $\{1,2,3\}$, and so on. The special property that we demand of the morphisms in the context of counting is that they should be one-to-one correspondences. For example, if you have a flock of exactly three sheep for your source, a representation of that flock *must* have $\{1,2,3\}$ as its target. Thus, the "essential nature" of the source that is preserved by the morphism, *in this context*, is the number of elements it contains.

There are a lot of possible morphisms—$n!$ to be exact, where n is the number of elements in the source and target.[5] When we are counting the number of elements in a set, we do not actually care about *which* morphism we grasp onto. But there is no choice about the target: it is $\{1,2,3,\ldots,n\}$ if and only if n is the number of elements in our source.

[5]The notation $n!$, pronounced "n factorial," means the product of all of the numbers from 1 through n. For example, 5! is $1\cdot 2\cdot 3\cdot 4\cdot 5 = 120$. For an exercise, you can find the six possible one-to-one correspondences from the set {red, blue, green} to the set $\{1,2,3\}$.

We could alter the counting process, and stipulate that a morphism be a one-to-one correspondence from the source to a *subset* of the target. But that would allow us to count the three oranges on our desk as "19, 3, 55," for example, which is useless.

Or is it? If that is our count, then we know that there are three oranges, because $\{19, 3, 55\}$ is a set of three numbers. But how do we know how many numbers are in the set $\{19, 3, 55\}$? We still have to count them, so this technique has not helped us.

Suppose that we require the count to go in order of size. Then the above example is invalid, but "3, 19, 55" is valid. As always, knowing the last number in the count is the point. In this case, we would then know that the source has *at most* 55 objects. This leads to the concept of *less than or equal*. We could now generate the science of inequalities by using this kind of morphism.[6]

Summary

If A represents B, we have three things: two objects, A and B, which from now on will be sets, and the relation between them, which from now on will be a morphism. When A and B have some additional "structure"—e.g., they are *finite* sets, or *ordered* sets—and we restrict the possible morphisms from A to B to have something to do with that structure—e.g., morphisms must be one-to-one correspondences, or order-preserving functions—then the existence of a representation from A to B gives us some information about A in terms of the standard object B—e.g., we can find out how many elements are in A, or at most how many elements are in A. Another example of adding structure to a set, allowing a more profound study of that set, is given by the sets of permutations to which we will add a *group structure*; see chapter 3.

In this book we explore some very explicit examples of representations. The things we consider are always mathematical objects such as sets, groups, matrices, and functions between them. We

---

[6]For a nice discussion of counting and its extension to infinite sets, see *One, Two, Three . . . Infinity* (Gamow, 1989).

show you how this works in detail in one particular case that we develop throughout the book and that gets us to our goal: mod *p* linear representations of Galois groups. We explain how these representations help to clarify the general problem of solving systems of polynomial equations with integer coefficients, and how they can sometimes lead to definitive results in this area.

Besides the representations discussed in this book, there are many other kinds of representation theories used today in mathematics. Representation theory is often needed to formulate interesting problems, as well as to solve them.

Chapter 2

GROUPS

Road Map

Mathematical sets can be interesting and easy to define, but very hard to understand in detail. For example, the set of all prime numbers is easy to define but is the source of many unsolved problems in number theory. If a set can be endowed with extra structure, it can help our understanding. One very common kind of extra structure is a "composition law" that turns the set into a *group*. The definition of a group is the most common way mathematicians have of formalizing the concept of *symmetry*.

The concept of a group is necessary for the representation theory we are developing. All of our objects are groups of one kind or another. There is a lot known about groups in general and about special kinds of groups in particular, and we will be able to deploy all this knowledge when we can impose a group structure on the objects we want to learn about.

In this chapter, we state the definition of *group* and look in detail at a particular example, the group SO(3) of rotations of a sphere.[1] We chose this example because it is easy to visualize but complicated enough to give you the full flavor of "groupness."

[1] SO(3) stands for "special orthogonal group in three dimensions."

The Group of Rotations of a Sphere

A group is a set along with a rule that tells how to combine any two elements in the set to get another element in the set. We usually use the word *composition* to describe the act of combining two elements of the group to get a third.

We start our consideration of groups by thinking about a beautiful perfect sphere, one foot in radius, made of pure marble. Let it rest in a spherical container so it just fits exactly. Assume that we have a perfect map of the earth drawn on the sphere, so we can refer to points on the sphere by the corresponding latitude and longitude of points on the earth. We ignore the fact that the real earth is not a perfect sphere.

To mark the initial position of this sphere in its container, draw a red dot on the sphere and on the container at the North Pole, and draw circles on the sphere and on the container where the equator is. We can also put a green dot on the equator, both on the sphere and the container, marking the Greenwich meridian. Now, there is exactly one way to place the sphere in the container so that the North Pole dots match, the equators match, and the Greenwich dots match.

We have not defined any groups yet. This sphere and container are just the (idealized) physical set-up we need to define the group that is called SO(3). We will first define the set SO(3) by telling about its elements. An element g of SO(3) is a rotation of the sphere inside the container. If we rotate the sphere by g, it will come to a new position in the container, which we can see because the two dots and the equator will be somewhere else. This is true except in the case when g is the "neutral element" in the group (see below).

Now, here is a very important point: If we take the sphere out of the container, toss it around, show it to our friends, and then put it back carelessly into the container, it will be in a new position. It is always possible to move the sphere from its original position into this new position *by some rotation*.

For instance, g might be rotation about the North Pole by 30°.[2] (This is the rotation of the earth in any 2-hour period.) Another

[2] In describing rotations by numbers of degrees, we shall always mean *counterclockwise* as we look down from above, as if we were trying to unscrew a light bulb.

element, h, for example, might be the rotation of the sphere 77° about an axis through the center of the sphere and the point of the earth at 38.89664°N, 77.04327°W.[3]

The key point about a group is the combination of any two elements to get a third. In our example of SO(3) this works as follows: We define the "composition" of any two rotations to be the rotation that is the result of doing first one and then the other rotation. (The order must be specified.) For example, if g and h are the rotations described above, their composition, which is denoted $h \circ g$, is the rotation f defined as follows: Start the sphere in the original position, then do the g rotation and then the h rotation, without returning the sphere to its original position in between. In other words, first rotate the sphere 30° around the North Pole, and then 77° around the *original* axis that goes through 38.89664°N, 77.04327°W in the container.

The sphere is now in some position that can also be reached by starting at the original position and rotating approximately 98° about the axis from the center of the earth to the point at 50°N, 92°W. This rotation is f, the result of composing g and h. In symbols, $f = h \circ g$. Calculating f was rather difficult; we used trigonometry and matrices.

Notice that we read compositions from right to left. It is important to remember that. In our example, $g \circ h$ is a different rotation from $h \circ g$. In fact, $g \circ h$ is rotation by approximately 98° about an axis through the center of the sphere and through the point at approximately 50°N, 62°W on the surface of the sphere. The fact that for some rotations g and h, $g \circ h \neq h \circ g$ can be expressed in more abstract language by saying that "the composition law in the group SO(3) is not commutative."[4]

Our group SO(3) contains a "neutral" element, which we will call e. It is rotation about any axis by 0°. In other words, it is

[3]This is the approximate location of an obelisk honoring George Washington in Washington, D.C. Because an obelisk looks like the tip of an axis, it seems like a good choice around which to rotate the sphere.

[4]Note that $g \circ h$ and $h \circ g$ rotate by the same number of degrees! Can you prove that this must be so? Take a look at chapter 15 for more discussion of this fact. An elegant method for composing rotations using spherical geometry may be found in (Penrose, 2005, pp. 206–8).

doing nothing at all to the sphere. You may object that this is not a rotation; it is nothing at all. But you must give it the honorary position of a rotation in our group. Otherwise, we would have trouble composing certain pairs of rotations.

For example, let g be the rotation around the North Pole by 30° discussed above, and let k be the rotation around the North Pole by −30° (in other words, by 30° *clockwise* when looking down.) Then $k \circ g = e$, and we need e in the group if the combination of any two elements in the group is to yield an element in the group.

You can extend this example and see that for any rotation x in SO(3), no matter what it is, there is another rotation y that undoes x, so that $y \circ x = e$. Do you also see that $x \circ y = e$? We call y the "inverse" of x and write $y = x^{-1}$.

EXERCISE: Convince yourself that $e^{-1} = e$.

We now come to another very important point: Consider the rotation where we rotate the sphere 360° about the North Pole. This will bring it back exactly to its starting position. Is this the same rotation as the neutral rotation e? We apparently have a choice here. If we use the usual English definition of the word "rotation," we have to say it is different. After all, it requires a lot more work to rotate the heavy sphere one full turn, compared to sitting back and doing nothing at all.

But this choice is not so good, because then our definition of "composition" is no good. We said that to compose $h \circ g$, we do g and then h to see that the rotation f has the same effect as their net result. Now, we could always add 360 to the number of degrees of rotation prescribed by f and get an apparently "different" rotation with the same effect. That is no good. We must have $h \circ g$ be one, and only one, particular element of SO(3).

So we really have no choice. In the useful definition of a rotation (in this context), a rotation is determined by what *effect* the rotation has on the sphere, and not what we did to get there. Suppose that you rotate the sphere by 45° about the point 30°N, 50°W, then pull it out of the container, doing a somersault with the sphere tucked into your shirt, put it back in the container exactly how it was when you pulled it out, and then rotate it another 45° about the same axis. Although this exercise will tire you out, the net result is simply the *same* element of SO(3) that is described more simply by saying: rotation by 90° about the point 30°N, 50°W.

Notice how in mathematics we can choose the definitions to fit our purpose, in this case to make SO(3) into a group. But after we announce and agree on a definition, we have to stick to it until further notice.

This is all we need to say in order to define the group SO(3). As a *set*, it is the set of all rotations of the sphere, as defined above. As a *group*, it has the extra feature of composition: Given any two rotations x, y we compose them by defining $x \circ y$ as that single rotation that has the same effect as doing first y and then x.

The General Concept of "Group"

DEFINITION: A *group* G is a set with a composition defined on pairs of elements, as long as three axioms hold true:

1. There is a neutral element e in G, so that $x \circ e = e \circ x = x$ no matter what element of the group is substituted for x.[5]
2. For any element x of G, there is some element y in G so that $x \circ y = y \circ x = e$.
3. For any three elements x, y, and z in G, we have $(x \circ y) \circ z = x \circ (y \circ z)$.

In the second rule, it should be realized that y may possibly be x itself (e.g., if $x = e$), but usually it is a different element. In all cases, we write $y = x^{-1}$. Also, even though the order of group composition usually matters, in the case of inverses, it does not: $x \circ x^{-1} = x^{-1} \circ x = e$.

The third rule is called the *associative law*. It has to hold in all groups, as it does in SO(3). Why is it true in SO(3)? Because whether we write $(x \circ y) \circ z$ or $x \circ (y \circ z)$, we end up doing first the

[5]The letter e is traditionally used for the neutral element, probably because it begins the German word *einheit*, which means "identity."

rotation z, then the rotation y, and finally the rotation x. In many groups, the associative law is much tougher to check (such as the matrix groups discussed in chapter 11).

In our definition, we used the symbol "$\circ$" to stand for the composition law of elements in our group. We could have used any notation, such as $x + y$, $x \cdot y$, $x * y$, or even no symbol, as in xy. Each group has its own law of composition. It can be whatever we define it to be—addition, multiplication, whatever—as long as the three axioms in the definition hold true.

DEFINITION: If G is a group, the *group law* is the rule that tells how to combine two elements in the group to get the third. We will usually write this combination as $x \circ y$, but occasionally as $x + y$ or even xy.

We will encounter many more examples of groups and group laws as our journey continues.

In Praise of Mathematical Idealization

The group SO(3) is dear to our hearts. It is around us all the time. Every second of the day and night, the earth performs the same rotation in SO(3), if we ignore slight discrepancies in the speed of the earth's rotation, and if we ignore the fact that the earth is not perfectly spherical. It pays to ignore these things for our purposes, because if we included every bump on every log we would never see the forest for the trees. That is, we would not be able to see any patterns that repeat. Every pattern would be ever so slightly (or more!) different from every other. We could never see the regularities to abstract from the welter of reality. Mathematics would never be born.

Even if your interest lies in the realities—whether you are a physicist, or just an ordinary person going about your daily round of activities—you have to abstract mathematically. For example, you use a clock that goes around twice in 24 hours, even though an astronomical day is not exactly 24 hours long. The great advances of physics were made through idealized approximations, so that patterns could be observed, for example, the inclined plane of Galileo, or the thought experiments of Einstein.

Group theory gives us a mathematical way to deal with symmetry. But the actual earth has no exact symmetries: No rotation (except the neutral, do-nothing rotation) takes it into a position where it exactly coincides with itself. The idealized sphere, on the other hand, is very symmetrical. This is reflected in the fact that SO(3) has many different elements—all rotations, which smoothly turn the sphere within its container.

Digression: Lie Groups

Our idealized earth turns smoothly all the time at exactly the same rate. There should be a concept of "infinitesimal rotation"—how much the earth turns in just an instant of time. This would be an "infinitesimal" element in SO(3), something like an angular velocity around a certain axis. Any actual rotation should be buildable by integrating some infinitesimal rotation. This is the germ of the theory of Lie groups and Lie algebras, named for the Norwegian mathematician Sophus Lie (pronounced "lee"). Lie group theory is thus a kind of marriage between calculus and group theory. It has a place in number theory too, but that is beyond the scope of our book.

A group such as SO(3) that has infinitesimal generators is called a *continuous group*. The other main type of group is a called a *discrete group*. In the latter, there is no smooth path from one element of the group to another. An example of a discrete group is the set of integers, where we "compose" two integers by adding them together. The neutral element is 0 and the inverse of x is $-x$. The next type of group we will study, a permutation group, is also a discrete group.

Other examples of Lie groups include the set of rigid motions of space to itself, used in crystallography and Newtonian physics; the set of Lorentz transformations of flat spacetime, used in relativistic physics; the set of rotations of a circle, also called SO(2); and the set of real numbers with composition given by ordinary addition. (Composition in the first three of these examples is given by doing first one motion and then the second, similar to composition in SO(3).)

Lie groups turn up when we study a geometric object with a lot of symmetry, such as a sphere, a circle, or flat spacetime. Because there is so much symmetry, there are many functions from the object to itself that preserve the geometry, and these functions become the elements of the group. As we will see, discrete groups can be also used to keep track of symmetries.

Chapter 3

PERMUTATIONS

Road Map

In this chapter we define and discuss permutation groups. In addition to being useful examples of groups, they are essential for our later definition of Galois groups. Galois groups are permutation groups of a certain kind: They permute roots of polynomials.

The *abc* of Permutations

Our next type of group goes back to the idea of one-to-one correspondence.[1] We start with a finite set, for example, $\{a, b, c\}$. One thing they do in elementary school is to figure out all the possible ways of ordering this set. There are six ways: a, b, c; a, c, b; b, a, c; b, c, a; c, a, b; c, b, a. We can view any one of these orderings as the result of a one-to-one correspondence of this set with itself. For example, the third ordering can be viewed as the result of

$$a \to b$$

$$b \to a$$

$$c \to c.$$

[1]Remember that a one-to-one correspondence (defined in chapter 1) is a special kind of *function*. Perhaps we should call it a "one-to-one transformation," but that means something else in standard mathematical jargon, so we will stick with the traditional and slightly misleading terminology.

We obtained this result by lining up the letters a, b, c in the original order (vertically for ease of viewing) and lining them up in the new order in the next column.

We think of a one-to-one correspondence in a dynamic way, following the model of our group SO(3). The elements of that group were defined in terms of actions that we took—rotating the sphere—and taking note only of the end result of the action. Similarly, we consider a *permutation* of the set $\{a, b, c\}$ to be some action we perform on the letters a, b, and c, and the end result will be the letters a, b, and c in a new order.

For example, we define an action called "g": g sends a to b, b to a, and c to c. There are several ways to think of the action. One is as a box, with an input slot on top, and an output slot on the bottom. We call this the "g-box," because it performs the action of the permutation g. If we drop a into the slot, b comes out. If we drop b in, a comes out. If we drop c in, c comes out.

Note that the g-box represents a *function*. It is not itself a permutation; it is a box. But it tells us how to get a permutation: Drop in a, b, and c successively, and out comes b, a, and c successively. We say that the permutation is defined by this box. (If we just sit there dropping in a's all day long, we will not find out what permutation is defined by the box, but if we drop in a, b, and c successively, we will.) This box picture gives us a vivid way to explain how to combine two permutations to get a third, as follows.

Each of the six possible orders of the three letters a, b, c defines one of these boxes. So we have six boxes, and they correspond to the six elements of this set of permutations. We combine elements of this set via a composition law. For example, let g be the permutation defined above, and let h be the permutation defined by the order c, b, a, or, equivalently, by the one-to-one correspondence

$$a \to c$$

$$b \to b$$

$$c \to a.$$

Then $h \circ g$ is defined as follows: Take the g-box and put the h-box underneath. Put an a into the g-box: out will drop b, and the b will drop into the h-box, and out will drop b. So b is the net result of dropping in a into the $h \circ g$-box. So the one-to-one correspondence defined by $h \circ g$ so far is

$$a \to b.$$

Next we drop in b and see what happens: An a drops out of the g-box, enters the h-box, and out comes c. If we drop in c in the top, we will see a c falling from the first to the second box, and ultimately out will come a. So the net result of $h \circ g$ is the one-to-one correspondence

$$a \to b$$

$$b \to c$$

$$c \to a.$$

This then is the one-to-one correspondence defined by $h \circ g$. We see by reading off the second column that the ordering defined by $h \circ g$ is b, c, a.

We think of the permutation as something we can "do" to the letters a, b, and c. We "send" the letter a, for example, to the letter that its arrow points to. In this way, we think of a permutation as an *instruction set.* For example:

$$a \to b$$

$$b \to a$$

$$c \to c.$$

Such an instruction set or rule is called a *function*, as defined in chapter 1. It is often useful to give a function a name. We can name this one g, because we called the box that does the same action the "g-box." We use the notation $g(x)$ for the result of "doing" the function g to the element x. In our example, $g(a) = b$, $g(b) = a$ and $g(c) = c$. This obviates the need for arrows and makes it easy to compose permutations.

For example, let g and h be the permutations above. Then the composed permutation $h \circ g$ has the following effect: If x stands for any of the letters a, b, or c, then $(h \circ g)(x) = h(g(x))$. The parentheses on the right-hand side of the equation are telling us to first do g to x, and then to do h to the result. Notice that the notation forces us to interpret $h \circ g$ as the composition of two functions read from right to left.

EXERCISE: Compute the permutation $g \circ h$, as a diagram with arrows.

SOLUTION: We know that h is the function diagrammed by

$$\begin{aligned} & a \to c \\ (h): \quad & b \to b \\ & c \to a. \end{aligned}$$

To find $g \circ h$, we first do h and then g. This can be arranged diagrammatically as

$$\begin{aligned} & a \to c \to c \\ (g \circ h): \quad & b \to b \to a \\ & c \to a \to b. \end{aligned}$$

Then $g \circ h$ can be diagrammed by eliminating the middle column to obtain the net result:

$$\begin{aligned} & a \to c \\ (g \circ h): \quad & b \to a \\ & c \to b. \end{aligned}$$

Notice that $h \circ g$ is not equal to $g \circ h$, that is, they are different permutations. Recall that $h \circ g$ is given by the diagram

$$\begin{aligned} & a \to b \\ (h \circ g): \quad & b \to c \\ & c \to a. \end{aligned}$$

So just as with SO(3), the group of permutations of $\{a, b, c\}$ is not commutative.

Permutations in General

We can play this game with any set A. The *group of permutations* of A is the set of functions from A to itself that are one-to-one correspondences. Composition of permutations is defined just as above, as composition of functions.

The neutral permutation e in the group of permutations of A is called the *identity permutation*. By definition, $e(x) = x$ for every x in A. For example, if $A = \{a, b, c\}$, then e is the permutation diagrammed by

$$(e): \begin{array}{l} a \to a \\ b \to b \\ c \to c. \end{array}$$

To find the inverse of any permutation, we switch the left and right columns, leaving the arrows in place, and then, if we wish, we reorder the rows so that the left-hand column is in the standard order. In symbols, if f is a permutation of the set A, then f^{-1} is the permutation defined by $f^{-1}(x) = y$ if and only if $f(y) = x$. For example, if $g \circ h$ is the permutation of $\{a, b, c\}$ we computed in the exercise on page 24, so that $g \circ h$ is given by the diagram

$$(g \circ h): \begin{array}{l} a \to c \\ b \to a \\ c \to b, \end{array}$$

then $(g \circ h)^{-1}$ is given by the diagram

$$(g \circ h)^{-1}: \begin{array}{l} c \to a \\ a \to b \\ b \to c, \end{array}$$

and then we can rearrange the lines to give

$$(g \circ h)^{-1}: \begin{array}{l} a \to b \\ b \to c \\ c \to a. \end{array}$$

EXERCISE: Compare $(h \circ g)^{-1}$ and $g^{-1} \circ h^{-1}$ and confirm that they are equal.

SOLUTION: Both $(h \circ g)^{-1}$ and $g^{-1} \circ h^{-1}$ are equal to

$$\begin{array}{l} a \to c \\ b \to a \\ c \to b. \end{array}$$

The permutations of a given set A are always a group,[2] using composition to combine pairs of elements. We use the symbol Σ_A to denote this group, read "Sigma sub A." If A has more than two elements, then the composition in Σ_A does not obey the commutative law.

EXERCISE: If A has n elements, explain why Σ_A has $n!$ elements.

Cycles

One of the niftiest elementary things about a permutation is called its *cycle decomposition*. In order to give you an interesting example, we start with a permutation of the first 10 letters of the alphabet, $\{a, b, c, d, e, f, g, h, i, j\}$. We write the permutation vertically rather

[2] We have not yet checked the associative axiom. If h, g, and k are permutations of A, we must verify that $h \circ (g \circ k) = (h \circ g) \circ k$. This follows from the fact that h, g, and k are functions from A to A.

than horizontally, to take up less space. Here is our permutation:

$$\begin{array}{cccccccccc} a & b & c & d & e & f & g & h & i & j \\ \downarrow & \downarrow & \downarrow & \downarrow & \downarrow & \downarrow & \downarrow & \downarrow & \downarrow & \downarrow \\ b & e & a & i & h & d & c & f & j & g \end{array}.$$

We start with the letter a, and play the following game. The letter a gets mapped to the letter b, which in turn gets mapped to the letter e, which in turn gets mapped to the letter h, which in turn gets mapped to the letter f, which in turn gets mapped to the letter d, which in turn gets mapped to the letter i, which in turn gets mapped to the letter j, which in turn gets mapped to the letter g, which in turn gets mapped to the letter c, which in turn gets mapped to the letter a. If you think about it for a moment, you will realize that we had to come back to a eventually.

EXERCISE: Convince yourself that we had to return to the letter a eventually.

SOLUTION: Because there are only 10 letters involved, the first thing to notice is that *some* letter has to show up on the list twice if we play the game long enough. Call the first letter to show up twice x, and suppose that x is *not* the letter a. We get a picture like this one:
$a \to b \to \cdots \to y \to x \to \cdots \to z \to x$. Because the permutation is one-to-one, we are forced to conclude that the letter before each of the two x's is the same as well, contradicting the assumption that x is the first letter to show up twice. The only way out is if a is the first letter to appear twice.

Put the whole list, starting with a, on one long line:

$$a \to b \to e \to h \to f \to d \to i \to j \to g \to c \to a.$$

This is called a *cycle*, and it is customary (though confusing at first) to write it as (*abehfdijgc*). Notice that we omit the second a.

What would happen if we start at h instead of a? We would get

$$h \to f \to d \to i \to j \to g \to c \to a \to b \to e \to h$$

which can be written $(hfdijgcabe)$. The key thing is that $(abehfdijgc)$ and $(hfdijgcabe)$ refer to the *same* cycle.

This particular permutation is a special case, because it consists of only one cycle. That means that when we start at a and trace what happens, we go through *all* of the letters being permuted before we return to a.

We try a different permutation to see what else might happen:

$$\begin{array}{cccccccccc} a & b & c & d & e & f & g & h & i & j \\ \downarrow & \downarrow & \downarrow & \downarrow & \downarrow & \downarrow & \downarrow & \downarrow & \downarrow & \downarrow \\ d & c & i & a & h & b & e & f & j & g \end{array}.$$

Now, when we start at a, we see that a gets mapped to d, which gets mapped right back to a. We can write that as the cycle (ad). But this time there are other elements of the set still to consider. What happens to b? We see that b gets mapped to c, which gets mapped to i, which gets mapped to j, which gets mapped to g, which gets mapped to e, which gets mapped to h, which gets mapped to f, which gets mapped to b. So the cycle that starts at b looks like $(bcijgehf)$. We can summarize the entire permutation by writing it as $(ad)(bcijgehf)$. If you look at this and want to find out what happens to any letter, it gets mapped to the letter to the right of it. If instead of a letter to the right there is a right parenthesis, find the letter immediately after the preceding left parenthesis. So we can see that j gets mapped to g, and f gets mapped to b.

For our purposes, the interesting information about a permutation will sometimes just be the lengths of the cycles. This particular permutation breaks down into two cycles: one of length 8 and one of length 2. We will see later that this captures some of the critical information about the permutation, even though, of course, reducing the permutation to a few numbers cannot tell us everything we might like to know about it.

EXERCISE: Verify the following equality in $\Sigma_{\{a,b,c,d,e\}}$: $(ab)(cde) \circ (ae)(bc)(d) = (ac)(bde)$.

CAUTION: You should read the section on permutations several times if necessary, because it is the basic idea that makes Galois theory come to life. We have used $\Sigma_{\{a,b,c\}}$ as our basic example. You could play around with Σ_A where A has 1, 2, 4, or 100 elements. If A has 100 elements, do not try to write out all the elements of Σ_A. There are 100! of them, and it would take you almost 3×10^{150} years if you wrote out one element every second.

Digression: Mathematics and Society

Mathematics is like a game. It has rules, and to enjoy playing or watching it, you have to know and understand the rules. Mathematicians make up the rules as they go along. There is sometimes an extrinsic benefit, because mathematics is used for many practical things, starting with counting and telling time, and including theoretical physics and constructing computers. Even number theory has extrinsic benefits nowadays, coming mostly from the theory of codes and ciphers, but also in acoustics, radar, and other areas. But even when no applications are immediately apparent, playing the game can be satisfying.

Many people believe that all of mathematics has already been discovered and codified. Mathematicians (they think) do nothing except rearrange the material in different ways for different types of students. This seems to be the result of the cut-and-dried method of teaching mathematics in many high schools and universities. The facts are laid out in the cleanest logical order. Little attempt is made to show how someone once had to invent it all, at first in a confused way, and that only later was it possible to give it this neat form. Many textbooks make no effort to tell about directions that are still to be explored, conjectures that are unproven, nor, of course, of ideas that are yet to be formulated.

Project yourself back in time to 1000 BC. Very little mathematics was known then. It all lay in the future to be discovered, debated, arranged, and improved. The situation today is nearly the same!

Very little has been discovered until now compared with the amount yet to come.

If you read books such as this one or the articles in the newspapers about mathematics, you know that new discoveries, sometimes very important ones, are made from time to time. But newspapers cannot report on the entire web of ideas and proofs that is continually being extended by mathematicians around the world. The same is true, of course, in the other sciences.

Chapter 4

MODULAR ARITHMETIC

Road Map

In this chapter we start to develop some elementary number theory. Modular arithmetic is sometimes taught in elementary school as "clock arithmetic," and is of fundamental importance in all of number theory. Modular arithmetic gives us our first examples of number systems other than the usual ones, and also gives new examples of groups.

The role of modular arithmetic in solving equations will be introduced briefly at the end of the chapter. This role will continue, sometimes in surprising ways, as our journey continues.

Cyclical Time

Modular arithmetic was invented and given that name by the nineteenth-century German mathematician Carl Friedrich Gauss, but the basic concept must be far older. It washed into American grade schools on the wave called "The New Math" in the 1950s and 1960s, and may still be found in some schools. The basic idea was usually illustrated by a problem of the following sort:

EXERCISE: Today is Tuesday. What day of the week will it be 25 days from now?

The student is supposed to realize that there are 7 days in a week: She can write 25 as $7+7+7+4$, and then ignore all but the 4, and conclude that 25 days from now it will be Saturday, the same as 4 days from now.

Another typical problem concerned clocks:

EXERCISE: If an (analog) clock is now showing 8 o'clock, what time will it be showing 33 hours from now?

The student is supposed to realize that clocks repeat themselves every 12 hours.[1] So 33 hours from now is $12+12+9$ hours from now, which (for the clock) is the same as 9 hours from now. That would mean the clock shows 17 o'clock, but we have to drop another 12 and get 5 o'clock.

If this type of word problem were the only reason to study modular arithmetic, we would not bother with it here. Rather, these problems illustrate a powerful general concept that will be critically important later. We stick to the clock example and use it to illustrate some notation. We have decided that 17 o'clock is just a synonym for 5 o'clock, and in general we can ignore any multiples of 12 we run into. Computer scientists have a compact notation for this: They write $17\%12 = 5$. The notation "$a\%12$" means "divide a by 12 and compute the remainder."

There is a problem here: In dealing with clocks, we use the numbers 1 to 12, whereas remainders go from 0 to 11. In other words, $24\%12 = 0$. This convention is in fact followed in 24-hour time, which runs from 0:00 to 23:59.

Similarly, we can talk about minutes past the hour by dividing by 60 and computing the remainder. For example, $74\%60 = 14$; if the minute-hand of the clock is at 13 now, in 74 minutes it will be at $13+14=27$.

---

[1]Digital clocks do not change the problem that much; they repeat every 24 hours, assuming that they distinguish AM from PM.

Congruences

The symbol "%" is used by computer scientists, but mathematicians find it more convenient to study these concepts with a *binary relation*[2] called *congruence*, which is similar to equality. We define two numbers as *congruent modulo n* if they have the same remainder upon division by n. For example, we will call 23 and 45 congruent modulo 11, because each leaves the remainder of 1 when divided by 11. Computer scientists would write this as $23\%11 = 45\%11$. The notation invented by Gauss to express this is

$$23 \equiv 45 \pmod{11}. \tag{4.1}$$

DEFINITION: We write $a \equiv b \pmod{n}$ (read as "a is congruent to b modulo n") if $a\%n = b\%n$. Equivalently, we write $a \equiv b \pmod{n}$ if $a - b$ is a multiple of n. The number n here is the *modulus* of the congruence.

EXERCISE: Show that the two definitions really are equivalent. In other words, show that if $a\%n = b\%n$, then $a - b$ is a multiple of n, and also show that if $a - b$ is a multiple of n, then $a\%n = b\%n$.

The $\equiv$ symbol can be used in most of the ways that the $=$ symbol is used. If $a \equiv b \pmod{n}$ and $b \equiv c \pmod{n}$, then $a \equiv c \pmod{n}$; this is the analogue for congruences of Euclid's famous axiom, "Things equal to the same thing are equal to each other."

Here is another numerical example to help you understand the notation: $100 \equiv 23 \pmod{11}$ because both leave a remainder of 1 when divided by 11. From this and the congruence (4.1), we can conclude that $100 \equiv 45 \pmod{11}$ without having to check again that both leave the same remainder when divided by 11.

But we can also deal with unknowns: For instance, if $x \equiv y \pmod{11}$ then we can see that $x + 2 \equiv y + 13 \pmod{11}$,[3] and this

[2] A *binary relation* is a relation between two things of the same kind, in this case whole numbers.

[3] It is because $2 \equiv 13 \pmod{11}$; this is like adding equals to equals.

congruence remains true whatever x and y are. However, we cannot actually divide either side of the congruence by 11 to see what the remainders are, because we do not know what x and y actually are.

Negative numbers can be included in the game. We will always use a positive remainder (or 0), even if we are dividing into a negative number. So, for example, $(-23)\%11 = 10$ because 11 goes into -23 just -3 times, leaving a remainder of 10. In other words, $-23 = -3 \cdot 11 + 10$.

What are all the positive numbers satisfying $x \equiv 5 \pmod 7$? They are 5, 12, 19, 26, 33, 40, ... (just keep adding 7). They make up what is called an "arithmetic progression."

We need to pause here and define *integer*, *prime number*, and *composite number*. We will be using these concepts all the time, starting in the next paragraph.

DEFINITION: An *integer* is a whole number—positive, negative, or zero. A *prime number* is a positive integer greater than 1 that has no positive divisors except itself and 1. A *composite number* is a positive integer greater than 1 that is not prime. We will also refer to a negative number as prime (respectively, composite) if its absolute value[4] is prime (respectively, composite).

The first few primes are 2, 3, 5, 7, 11, 13, 17, 19, and 23. There are infinitely many prime numbers. Some composite numbers are 4, -52 and 1000.

A famous theorem of the nineteenth-century mathematician Lejeune Dirichlet says that an arithmetic progression, such as 5, 12, 19, 26, 33, 40, ... contains an infinite number of prime numbers, as long as there is no common divisor greater than 1 of all the numbers in it. (For example, the arithmetic progression 2, 6, 10, 14, ... cannot contain infinitely many prime numbers because all of the numbers in the progression have a common divisor of 2.)

It may have occurred to you that we left negative numbers out of the arithmetic progression. If so, you are right. Besides 5, 12,

[4] If a is a real number, then $|a|$ denotes its absolute value: $|a| = a$ if $a \geq 0$, and $|a| = -a$ if $a < 0$.

19, ..., we should include -2, -9, -16, -23, ... (just keep subtracting 7) in the list of all numbers $\equiv 5 \pmod 7$.

It does not make sense to multiply hours, but we will use congruence notation for multiplication (we will get to division later). It takes a bit of algebra to show that if $a \equiv b \pmod{12}$ and $c \equiv d \pmod{12}$, then $a + c \equiv b + d \pmod{12}$ and $ac \equiv bd \pmod{12}$. For example, $3 \equiv 15 \pmod{12}$ and $8 \equiv 80 \pmod{12}$, so $3 + 8 \equiv 15 + 80 \pmod{12}$, and $3 \cdot 8 \equiv 15 \cdot 80 \pmod{12}$. In other words, $11 \equiv 95 \pmod{12}$ and $24 \equiv 1200 \pmod{12}$.

There is nothing magical about the number 12 here being used as modulus. For example, we have $3 \equiv 13 \pmod 5$ and $13 \equiv 8 \pmod 5$, so it follows that $3 \equiv 8 \pmod 5$. However, although we know that $3 \equiv 20 \pmod{17}$, we cannot combine this last congruence with any one of the others in the preceding sentence, because they have different moduli. We can only add, subtract, and multiply congruences if they have the *same* modulus: We have $2 \equiv 7 \pmod 5$ and $3 \equiv 13 \pmod 5$, and so $2 \cdot 3 \equiv 7 \cdot 13 \pmod 5$.

EXERCISE: Show that if $a \equiv b \pmod n$ and $c \equiv d \pmod n$, then $ac \equiv bd \pmod n$.

SOLUTION: The hypotheses tell us that $a - b$ and $c - d$ are multiples of n. Therefore $ac - bd = a(c - d) + d(a - b)$ is also a multiple of n.

What about division? Here is where we can run into trouble. We know that with the usual equals sign, if $ac = bc$, and $c \neq 0$, then $a = b$; in other words, we can divide both sides of an equality by any number except 0, and get a correct answer. In modular arithmetic, sometimes this works and sometimes it does not, so you cannot count on it. We try this modulo 12: We have $1 \cdot 7 \equiv 13 \cdot 7 \pmod{12}$, and sure enough, $1 \equiv 13 \pmod{12}$. We try another example: $3 \cdot 4 \equiv 6 \cdot 4 \pmod{12}$—but trouble here because $3 \not\equiv 6 \pmod{12}$.

The problem here is that 12 is a composite number. In fact, something similar can always happen whenever the modulus is a composite number. What if we stick to a prime modulus? Then we are in good shape: We have a cancellation law if the modulus is prime.

At first glance, you might think that there is still a problem. After all, we can write $1 \cdot 5 \equiv 0 \cdot 5 \pmod 5$, but if we cancel the 5's, we get $1 \equiv 0 \pmod 5$. But this is not really a problem with our definition of arithmetic modulo 5; we tried cancelling 5's, and we were ignoring the fact that $5 \equiv 0 \pmod 5$. The rule is that we can only cancel *nonzero* common factors, and "zero" does not mean *equal* to 0; it means *congruent* to 0 when we are working with congruences.

Arithmetic Modulo a Prime

There is a special symbol used when working with the integers modulo p (where p is any prime): $\mathbf{F}_p$. The letter "**F**" here stands for *field*.

> **DEFINITION**: A *field* is a number system[5] where we can divide by anything nonzero.

$\mathbf{F}_p$ is defined as the set $\{0, 1, \ldots, p-1\}$ with addition and multiplication defined as follows: If x, y, and z are in $\mathbf{F}_p$, $x + y = z$ in $\mathbf{F}_p$ exactly when $x + y \equiv z \pmod p$ and $xy = z$ in $\mathbf{F}_p$ exactly when $xy \equiv z \pmod p$. In other words, we allow ourselves to use equality signs in $\mathbf{F}_p$ where we would use congruence signs among integers.

We say that $\mathbf{F}_p$ is a "field with p elements." We also say that $\mathbf{F}_p$ is a number system with *characteristic* p.[6] It is far from obvious that $\mathbf{F}_p$ is a field, because we need to rethink our usual definition of division. $\mathbf{F}_p$ does *not* contain fractions; rather, if x is any nonzero element of $\mathbf{F}_p$, then there is some element y in $\mathbf{F}_p$ so that $xy = 1$. (In terms of congruences, this means that $xy \equiv 1 \pmod p$.) Then to "divide" by x, we instead multiply by y. In order for the number y always to exist, it is essential that we use a *prime* modulus.

---

[5]We are deliberately leaving the concept of "number system" vague. In this book, a number system is a set made out of numbers so that all of the usual rules of algebra hold, such as the commutative and associative laws of addition and multiplication, the distributive law of multiplication over addition, and so on.

[6]A field has characteristic p if $\underbrace{a + a + \cdots + a}_{p \text{ times}} = 0$ for any element a in the field.

We can write out the addition and multiplication tables for any specific prime p. We use the numerals from 0 to $p-1$ for the elements of the field.

For example, addition and multiplication in $\mathbf{F}_2$ look like this:

+	0	1
0	0	1
1	1	0

×	0	1
0	0	0
1	0	1

The first table means that $0+0 \equiv 0 \pmod 2$, $0+1 \equiv 1 \pmod 2$, $1+0 \equiv 1 \pmod 2$, and $1+1 \equiv 0 \pmod 2$. This is the same as saying about whole numbers:

- Even plus even is even.
- Even plus odd is odd.
- Odd plus even is odd.
- Odd plus odd is even.

The second table means that $0 \cdot 0 \equiv 0 \pmod 2$, $0 \cdot 1 \equiv 0 \pmod 2$, $1 \cdot 0 \equiv 0 \pmod 2$, $1 \cdot 1 \equiv 1 \pmod 2$. This is the same as saying about whole numbers:

- Even times even is even.
- Even times odd is even.
- Odd times even is even.
- Odd times odd is odd.

As you probably know, computers use this binary arithmetic in their innermost guts (or brains, depending on how you think of computers.)

EXERCISE: Write out the addition and multiplication tables for $\mathbf{F}_5$. Remember that every integer outside of the range 0–4 must be replaced by its remainder modulo 5.

SOLUTION: We have

+	0	1	2	3	4
0	0	1	2	3	4
1	1	2	3	4	0
2	2	3	4	0	1
3	3	4	0	1	2
4	4	0	1	2	3

×	0	1	2	3	4
0	0	0	0	0	0
1	0	1	2	3	4
2	0	2	4	1	3
3	0	3	1	4	2
4	0	4	3	2	1

For example, to explain the entry in the second table for the spot in the row labeled 3 and the column labeled 4, we multiply $3 \cdot 4$, getting 12, and divide by 5, getting a remainder of 2. Then we know that $3 \cdot 4 \equiv 2 \pmod 5$, so we put a 2 there.

Notice that the pattern in the addition table is easy to spot, sort of like a wave going from the upper left-hand corner, along the diagonals, down to the lower right corner. Or if you look just at the columns, from left to right, they might make you think of a barber pole turning.

The patterns in the multiplication table are less obvious. There is a border of 0's, because 0 times anything is 0. There is a symmetry around the main diagonal (the diagonal that goes from the upper left-hand corner to the lower right-hand corner). That is because multiplication is commutative: $a \cdot b = b \cdot a$. There is also a symmetry about the other diagonal if you strip off the border of 0's. Look at the multiplication table of a larger prime, say $p = 13$.

$\times$	0	1	2	3	4	5	6	7	8	9	10	11	12
0	0	0	0	0	0	0	0	0	0	0	0	0	0
1	0	1	2	3	4	5	6	7	8	9	10	11	12
2	0	2	4	6	8	10	12	1	3	5	7	9	11
3	0	3	6	9	12	2	5	8	11	1	4	7	10
4	0	4	8	12	3	7	11	2	6	10	1	5	9
5	0	5	10	2	7	12	4	9	1	6	11	3	8
6	0	6	12	5	11	4	10	3	9	2	8	1	7
7	0	7	1	8	2	9	3	10	4	11	5	12	6
8	0	8	3	11	6	1	9	4	12	7	2	10	5
9	0	9	5	1	10	6	2	11	7	3	12	8	4
10	0	10	7	4	1	11	8	5	2	12	9	6	3
11	0	11	9	7	5	3	1	12	10	8	6	4	2
12	0	12	11	10	9	8	7	6	5	4	3	2	1

Can you see any obvious patterns besides the border of 0's and these two symmetries? We do not.

Incidentally, we worked out this multiplication table by working across each row, adding a certain constant (namely, the heading of the row) and reducing modulo 13 when necessary. For example, we worked out the row numbered 4 by starting with $0, 4, 8, 12, 16, \ldots$, and then reducing modulo 13. Because of the symmetry of the table, after we had done the seventh row we could then get the others without any more work. The last row is just the reverse of the second row, for example.

There are fields that contain a composite number of elements. The number of elements in any finite field is always a power of a prime. Fields with composite numbers of elements are relevant to the type of number theory we are exploring, but they are too difficult for the level of this chapter. We will not need them in this book, but it is worth mentioning that they can be defined.

Modular Arithmetic and Group Theory

One thing mathematicians do is connect concepts that occur in different trains of thought. In the first few chapters, we were concerned with groups. In this chapter, we have discussed modular arithmetic, which is our way of dealing with divisibility of numbers and remainders. We can connect these concepts by noticing that modular arithmetic gives us not one but two groups for each prime p:

1. $\mathbf{F}_p$ is a group under addition. The neutral element is 0. The inverse of any number $a \pmod p$ is just $p - a \pmod p$.
2. If we throw 0 away from $\mathbf{F}_p$, we get a new set, with only $p - 1$ elements, called $\mathbf{F}_p^\times$, pronounced "eff-pea-cross." It is a group under multiplication. The neutral element is 1. Every element a has an inverse. Because $\mathbf{F}_p$ is a field, to find the inverse of a, you solve $ax \equiv 1 \pmod p$ by "dividing" both sides by a.

For example, to find the inverse of 3 in $\mathbf{F}_7^\times$, we have to solve $3x \equiv 1 \pmod 7$. Trial-and-error shows that $x = 5$. So $3^{-1} = 5$ in $\mathbf{F}_7^\times$. Solving this congruence for large moduli without using trial-and-error requires a tool called the Euclidean algorithm, which we will not need in this book.

No matter what the prime p is, these two groups are "commutative"—that is, the group law satisfies $x+y=y+x$ in the first case and $x \cdot y = y \cdot x$ in the second. So these groups are different from, and rather simpler than, SO(3) and Σ_n.[7]

The groups $\mathbf{F}_p$ and $\mathbf{F}_p^\times$ (no matter what p is) also share a strong property that again fails to hold for SO(3) and for Σ_n if $n > 2$. They are *cyclic*.

DEFINITION: A group is *cyclic* if it contains some element, call it g, so that we can get every non-neutral element of the group by repeating the group composition on g or g^{-1}. This particular element g is called a *generator* of the group.

The group $\mathbf{F}_p$ is cyclic because starting with $g = 1$, we get 1, $1+1 \equiv 2$, $1+1+1 \equiv 3, \ldots$, until we get all the elements of $\mathbf{F}_p$.

The group of ordinary integers with group operation addition (and neutral element 0) is also cyclic, generated by $g = 1$, because any positive integer is a sum of 1's, and any negative integer is a sum of -1's.

The fact that $\mathbf{F}_p^\times$ is cyclic is very interesting but would take us too long to prove here. You can check it for any particular prime p. For instance, if $p = 5$, starting with $g = 2$, we get 2, $2 \cdot 2 \equiv 4 \pmod 5$, $2 \cdot 2 \cdot 2 \equiv 3 \pmod 5$, and $2 \cdot 2 \cdot 2 \cdot 2 \equiv 1 \pmod 5$.

EXERCISE: Show that $\mathbf{F}_{17}^\times$ is cyclic.

SOLUTION: We use trial-and-error. We start by seeing if 2 might be a generator of the group: $2^1 \equiv 2$, $2^2 \equiv 4$, $2^3 \equiv 8$, $2^4 \equiv 16$, $2^5 \equiv 15$, $2^6 \equiv 13$, $2^7 \equiv 9$, and $2^8 \equiv 1$, and now the pattern repeats: $2^9 \equiv 2$, $2^{10} \equiv 4$, etc. So we do not encounter every element of $\mathbf{F}_{17}^\times$ by taking powers of 2.

Next, we try 3: $3^1 \equiv 3$, $3^2 \equiv 9$, $3^3 \equiv 10$, $3^4 \equiv 13$, $3^5 \equiv 5$, $3^6 \equiv 15$, $3^7 \equiv 11$, $3^8 \equiv 16$, $3^9 \equiv 14$, $3^{10} \equiv 8$, $3^{11} \equiv 7$, $3^{12} \equiv 4$, $3^{13} \equiv 12$, $3^{14} \equiv 2$, $3^{15} \equiv 6$, and $3^{16} \equiv 1$. This shows that 3 is a generator of $\mathbf{F}_{17}^\times$.

---

[7] Σ_n is short for the permutation group $\Sigma_{\{1,2,3,\ldots,n\}}$.

Modular Arithmetic and Solutions of Equations

We can use congruences to prove that some equations have no integral solutions. For example, suppose we want to show that there are no integers x and y with the property that $x^2 + y^2 = 11$. Suppose that there were such x and y. Now if two sides of an equation are equal integers, they are of course congruent modulo any modulus we like (because their difference is 0, and 0 is divisible by any positive number.)

So if $x^2 + y^2 = 11$, then $x^2 + y^2 \equiv 11$ modulo any number we like. Let's choose 4. Then $x^2 + y^2 \equiv 11 \pmod 4$. We will now see that this is impossible. Why?

Consider the squares: they are 0, 1, 4, 9, 16, 25, 36, 49, ... Modulo 4, they are 0, 1, 0, 1, 0, 1, 0, 1, ... (Do you see why there is such a simple pattern?)[8] So x^2 is either 0 or 1 modulo 4. By the same token, y^2 is either 0 or 1 modulo 4. Therefore, their sum must either be $0 + 0 \equiv 0$, $0 + 1 \equiv 1$ or $1 + 1 = 2$ modulo 4. OK, what's 11? It is 3 modulo 4 and so cannot be a sum of two squares.

You might say: "Big deal. To do the original problem, all I needed to do was to test all positive x's and y's less than $\sqrt{11}$, which does not take long." But now, we can show that there are no integers x and y with the property that $x^2 + y^2 = 4444444444444411$. Our method works just as before, while your method would take some time. Also, we get a general theorem: if $a \equiv 3 \pmod 4$, then there are no integers x and y with the property that $x^2 + y^2 = a$.

---

[8]This is because if n is even, so is n^2, while if $n = 2k + 1$ is odd, then $n^2 = (2k + 1)^2 = 4k^2 + 4k + 1$ leaves a remainder of 1 when divided by 4.

Chapter 5

COMPLEX NUMBERS

Road Map

We introduce (or review) the complex numbers, an extension of the real numbers useful for solving equations. The set of complex numbers is another example of a field. It is handy because every polynomial in one variable with integer coefficients can be factored into linear factors if we use complex numbers. Equivalently, every such polynomial has a complex root. This gives us a standard place to keep track of the solutions to polynomial equations.

As with the finite fields of chapter 4, we will be working with complex numbers in most of the chapters to follow. In this chapter, we also introduce an important subset of the complex numbers, namely, the set of all "algebraic numbers"—those numbers that are roots of polynomial equations with integer coefficients. This set is also a field, and will be important when we study the structure of solutions of polynomial equations.

Overture to Complex Numbers

We do not absolutely need to use complex numbers to solve polynomial equations with integer coefficients. Instead, we can use a complicated algebraic recipe for inventing solutions as we need them and then keep track of them as we continue to add more solutions. If we wish to ignore the complex numbers, we can simply assume the existence of a large number system that contains all the solutions to all equations of the form $f(x) = a_n x^n + a_{n-1}x^{n-1} + \cdots + a_1 x + a_0 = 0$, where the coefficients $a_n, a_{n-1}, \ldots, a_1, a_0$ are integers. In this case, you must remember that the same root can occur for many different polynomials. A sophisticated method is required for keeping track of these solutions, but it can be done.

One advantage of working with complex numbers is that each solution comes with its own personality. For example, the solutions of the equation of $x^2 + 1 = 0$ are $i = \sqrt{-1}$ and $-i = -\sqrt{-1}$. These two complex numbers have all the same *algebraic* properties, but they are not equal to each other—they are "twins." (For more about this mystery, you can refer to *Imagining Numbers* (Mazur, 2003).) But if we have the complex numbers sitting before us, we can call one of them i and the other $-i$. How, you might ask, do you tell i and $-i$ apart? Easy. Multiplication by i rotates the complex plane[1] by 90° *counterclockwise* as we look down upon it, while $-i$ rotates it the same amount *clockwise*.

Now, we define complex numbers, starting with the real numbers.

DEFINITION: A *real number* is any number that can be expressed as a decimal.

For example, $0 = 0.0, 1 = 1.0$, $\frac{3}{5} = 0.6$, and $-\sqrt{2} = -1.4121356\ldots$ are real numbers. A real number can be expressed as a *terminating* or *repeating* decimal if and only if it is the ratio of two integers. The set of all real numbers is usually denoted by the symbol **R**.

Complex numbers were forced upon the world when mathematicians began solving cubic[2] equations. You may think that complex numbers would have shown up as soon as someone tried to solve the quadratic equation $x^2 + 1 = 0$, but in fact mathematicians were quite happy to proclaim that this equation simply had no solutions. After all, the equation $0x = 1$ has no solutions, so why should it be a problem if $x^2 + 1 = 0$ has no solutions?

---

[1]The *complex plane* is the (x, y)-plane on which we plot the complex number $x + iy$ as the point (x, y). See below for how to multiply complex numbers.

[2]Degree 3: For example, $x^3 + x - 1 = 0$.

The situation with cubic equations was more complicated. Mathematicians discovered a formula for solving cubic equations similar to but more complicated than the quadratic formula $\frac{-b\pm\sqrt{b^2-4ac}}{2a}$ which solves the quadratic equation $ax^2+bx+c=0$. In some cases, however, some of the numbers in the cubic formula were square roots of negative numbers, even though in the end all of the solutions to the cubic equation were real numbers. We are not going to go through the algebra, but here is a very explicit example. The equation $x^3-7x+6=0$ has the three solutions $x=1$, $x=2$, and $x=-3$. If you try to solve this equation by using the cubic formula (which is like the quadratic formula, only much more complicated), along the way you unavoidably encounter the square root of a negative number.

Complex Arithmetic

Whether or not we believe that complex numbers really exist, we need to come up with rules for manipulating them. And the rules are simple enough.

DEFINITION: A *complex number* is a number of the form $a+bi$, where a and b are real numbers.

So $3+4i$ and $\sqrt{2}+\pi i$ are complex numbers;[3] so is $0+0i$ (which is another way of writing 0). Sometimes, we write the "i" before the second number: $2-i\sin 2$ is also a complex number.[4]

In the complex number $a+bi$, we call a the *real part* and bi the *imaginary part*. (Actually, mathematicians normally reserve the term "imaginary part" for the real number b, but that is a bit confusing, and we will not follow that usage.) We consider every real number x to be a complex number too, but write x instead of $x+0i$.

[3]The symbol π stands for the area of a circle of radius 1, which is approximately 3.14159265. It is a nonrepeating infinite decimal.

[4]If b is negative, we can write $a-(-b)i$ rather than $a+bi$.

Now we need to know how to do arithmetic with complex numbers. Addition is simple: we just add the real and the imaginary parts separately. In other words, $(a + bi) + (c + di) = (a + c) + (b + d)i$.

EXERCISE: What is the sum of $2 + 3i$ and $4 - 7i$?

SOLUTION: We add each part, and get
$(2 + 3i) + (4 - 7i) = (2 + 4) + (3 - 7)i = 6 - 4i$.

Subtraction is done similarly; each piece is subtracted separately. Formally, we write $(a + bi) - (c + di) = (a - c) + (b - d)i$.

EXERCISE: Subtract $4 - 7i$ from $2 + 3i$.

SOLUTION: We subtract each part, and get
$(2 + 3i) - (4 - 7i) = (2 - 4) + (3 - (-7))i = -2 + 10i$.

Multiplication is trickier. We use the distributive law of multiplication over addition (that is the one that says $A(B + C) = AB + AC$) and whenever we get $i \cdot i$ (which is also written i^2), we replace it by -1. In other words, we adopt the convention that $i^2 = -1$. Formally, the rule looks like

$$(a + bi)(c + di) = (ac - bd) + (ad + bc)i,$$

because we get the term $(bi)(di) = bdi^2 = -bd$.

EXERCISE: What is the product of $2 + 3i$ and $4 - 7i$?

SOLUTION: Use the formula to get
$(2 + 3i)(4 - 7i) = (2 \cdot 4 - 3 \cdot (-7)) + (2 \cdot (-7) + 3 \cdot 4)i = 29 - 2i$.

Finally, there is division. Rather than tell how to divide one complex number by another, we give a formula for the reciprocal of a complex number. Then, just like with division of fractions, rather than dividing by a complex number, you can multiply by the reciprocal.

There is a short computation to do before we tell you what the reciprocal is. That is to multiply $a+bi$ by $a-bi$. We get

$$(a+bi)(a-bi)=a^2+b^2. \tag{5.1}$$

Notice that a^2+b^2 is never 0 unless both a and b are 0 (and hence $a+bi=0$). Notice also that a^2+b^2 is always a real number.

EXERCISE: Check equation (5.1).

DEFINITION: The *complex conjugate* of the complex number $a+bi$ is the complex number $a-bi$.

Here is how we get the reciprocal of the number $a+bi$: We multiply the numerator and denominator by $a-bi$. This gives:

$$\frac{1}{a+bi}=\frac{1}{a+bi}\cdot\frac{a-bi}{a-bi}=\frac{a-bi}{a^2+b^2}=\frac{a}{a^2+b^2}-\frac{b}{a^2+b^2}i.$$

So the inverse of $a+bi$ is the rather complicated complex number $\frac{a}{a^2+b^2}-\frac{b}{a^2+b^2}i$. The inverse of $a+bi$ can be written just as $\frac{1}{a+bi}$, but it helps to put all complex numbers into standard $x+yi$ form, so we can compare them.

EXERCISE: What is the inverse of $3+4i$? Write your answer in the form $x+yi$.

SOLUTION: We use the formula, and get $\frac{1}{3+4i}=\frac{3}{3^2+4^2}-\frac{4}{3^2+4^2}i=\frac{3}{25}-\frac{4}{25}i$.

Now, to divide complex numbers, we multiply the dividend by the reciprocal of the divisor.

EXERCISE: What is $\frac{1+2i}{3+4i}$?

SOLUTION: We use the solution to the previous exercise, and get

$$\frac{1+2i}{3+4i}=(1+2i)\left(\frac{1}{3+4i}\right)=(1+2i)\left(\frac{3}{25}-\frac{4}{25}i\right)=\frac{11}{25}+\frac{2}{25}i.$$

Complex Numbers and Solving Equations

What is the point? First, mathematicians proved that using the set of complex numbers, which we call **C**, will not lead to any contradictions. Even though you might not like the idea of a square root of -1, pretending that it exists will not get you into logical trouble.

Second, and this is really the key point, the job of adding on square roots (or cube roots, or anything else that you might like to add on) stops with **C**. You do not need to add on some other new symbol j with $j^2 = i$, for example; there already is a square root of i in **C**.

EXERCISE: Let $\alpha = \frac{1}{\sqrt{2}} + \frac{1}{\sqrt{2}}i$. Check that $\alpha^2 = i$.

In fact, every root of a complex number is already in **C**. But even more than that is true:

THEOREM 5.2: Let $f(x)$ be a polynomial whose coefficients are any complex numbers. (For example, $f(x)$ might have integer coefficients.) Then the equation $f(x) = 0$ has solutions in **C**.

Digression: Theorem

A *theorem* is a mathematical statement that can be proved to be true. Despite the similarity to the word "theory," there is nothing speculative about a theorem.

Modern mathematical usage distinguishes between various terms, all of which refer to mathematical statements that are proved. The word *lemma* refers to a minor "helping" statement used along the way toward proving a theorem. A *corollary* is an immediate and easy consequence of a theorem.

Algebraic Closure

The fancy way of conveying the content of Theorem 5.2 is to say that **C** is *algebraically closed*. The first really good proof was given by Gauss, though earlier mathematicians such as Euler knew this fact as well.[5]

We are not going to need to use all of **C** very often. It contains numbers such as π, which cannot be understood algebraically. We will restrict ourselves mostly to that part of **C** which consists of the solutions of all equations $f(x) = 0$ where $f(x)$ is a polynomial with *integer* coefficients. That set is called $\mathbf{Q}^{\text{alg}}$, and it exists as a subset of **C**. In fact, $\mathbf{Q}^{\text{alg}}$ itself is a field! Sample elements of $\mathbf{Q}^{\text{alg}}$: 0, 1, 3, $-\frac{1}{4}$, $\sqrt{11}$, i, $\sqrt[3]{2} - 3i$, and the solutions in **C** of the equation $x^5 + x - 1 = 0$.

---

[5] Consult (Nahin, 1998) for a historical summary.

Chapter 6

EQUATIONS AND VARIETIES

Road Map

We now come face to face with the motivating problem of this book: equations and how to solve them. Although as number theorists we are primarily interested in integral or fractional solutions of systems of polynomial equations with integral coefficients ("Diophantine equations"), it will be helpful to consider also

- equations and their solutions in general, and
- solutions to equations involving number systems other than the integers.

In general, we cannot simply list all of the solutions of a given system of equations. Often, we cannot find all of them, or there are infinitely many solutions, or there are no solutions—but we may not know that (yet). Even if we can list all the solutions to some system of equations, such a list does not give us full understanding of the *structure* of the set of solutions. In this chapter we broach the question: "What are interesting structural properties of sets of solutions?"

Also in this chapter, we define *variety*, a concept that gives a concise way to discuss solutions in varying number systems to a fixed system of polynomial equations.

The Logic of Equality

An *equation* is a statement, or assertion, that one thing is identical to another. For example, "The first president of the United States was George Washington." Equality is timeless: Although we wrote "was," because both sides of the equation existed in the past, logically speaking, "is" or "will be" have the same force as "was."

Philosophically, equality can get pretty complicated. Consider the definition: "A unicorn is a horse with a single horn." A unicorn and a single-horned horse are identical—by definition—even though they (it) do not exist. Yet "is" might imply existence.

To rid ourselves of this problem, we drop the word "is" and its various forms, and replace it with the symbol =. Now, we do not have to take an ontological stand on the terms on either side of the symbol. Whatever their mode of being, we are asserting their identity.

We may also use symbols, usually single letters, to stand for the terms. For example, if F stands for the first president of the United States and G stands for George Washington, then the assertion in the first paragraph can be symbolized as follows: $F = G$.

Before algebra was invented, this kind of symbolism was unavailable, and people had to use ordinary language, which was often cumbersome. Nevertheless, equations were important even then, and have continued to become more and more useful as science has advanced.

Almost anything we do can be phrased as an equation. For example, solving $X = D$ for the unknown X, where D stands for "what we are having for dinner tonight," is equivalent to finding out what we are having for dinner tonight. From now on, though, we will restrict our discussion to mathematical equations, where the terms denote mathematical entities.

The History of Equations

Long before algebra as we know it, ancient peoples were working with equations:

1. The base angles of an isosceles triangle are equal.

2. If a circular bath has diameter 10 cubits, then the circumference is 30 cubits.
3. A triangle whose three sides have lengths 3, 4, and 5 is a right triangle.

The first equation is a proposition from Euclid. The assertion is claimed to hold for all isosceles triangles. So it is a general theorem.

The second equation is from the Bible, describing a large basin in the Temple in Jerusalem. It is not exactly true, but no doubt it was meant to be only approximate. In this book, we will be concerned only with exact equations.

The third equation is an example of a solution to what is called a *Diophantine equation* because it has unknowns (the three side-lengths) which are constrained to be integers (whole numbers). The equation we are thinking of here is $x^2 + y^2 = z^2$, which expresses the Pythagorean Theorem about the side lengths x, y, and z of a right triangle. As a mere equation about any old right triangles, we can suppose x, y, and z are any positive (real) numbers. What makes it a Diophantine equation is our self-imposed desire to restrict the solutions we are interested in to whole numbers.

Diophantus was a Greek mathematician from postclassical times whose writings influenced the founders of modern European number theory. Because Diophantus lived before the invention of symbolic algebra, his work can be hard to read. It is amazing what ancient mathematicians accomplished without the aid of the slick symbolic language we use now.

When symbolic algebra was first set down in Europe in a treatise by François Viète in the late 1500s, he called it "the art of finding the solutions to all problems."[1] In short order, René Descartes added the connection between algebra and geometry, and he also thought that now all geometrical problems could be first quantified and then solved. Because he also thought that the physical universe was governed entirely by geometrical laws, he felt that "analytic geometry" or "algebraic geometry" could be used to discover all of science.

[1] You can find an English translation of this treatise at the end of (Klein, 1992).

Soon after Descartes, people began to follow his program—with many changes, of course. One goal was to discover new scientific truths. Isaac Newton and Gottfried Leibniz, among others, found that algebra did not suffice to solve all of their scientific problems—they had to invent calculus.[2] Another goal, initiated by Fermat, shortly after Descartes, was the revival of pure contemplation of equations among whole numbers just for fun, for curiosity, or for the greater glory of God. It is this branch of mathematics that contains the representation theory, the Galois theory, and the number theory we want to explore in this book.

Even the part of number theory devoted to the investigation of integers soon turned out to require the use of calculus and other parts of mathematics. But we will restrict ourselves in this book to ideas that can be explained using only algebra.

Z-Equations

In German, the word for number is *zahl* (which is cognate with our word "tell," as in "teller who counts money at the bank"). Therefore, it is customary to use the symbol **Z** to denote the set of all integers, namely $\{0, 1, -1, 2, -2, 3, -3, \dots\}$. We also need to consider fractions:

> **DEFINITION**: A *rational number* is any number that can be expressed as the ratio of two integers. For example, 0, 44, and $-\frac{3}{7}$ are rational numbers, whereas π and $\sqrt{2}$ are not. Real numbers that are not rational are called *irrational numbers*.

A real number is rational if and only if it is a terminating decimal or a repeating decimal. All other decimals are "irrational" real numbers. The set of all rational numbers is usually denoted by the symbol **Q**.

---

[2]You can read the history of physics and how it spurred the development of mathematics in books such as (Ball, 1960; Boyer, 1991; Smith, 1958; Struik, 1987).

In most of this book, we will deal with equations where all the constants are integers; for example,

$$y^2 = x^3 - 3x + 14. \tag{6.1}$$

The experts call these *equations defined over the integers*. We will call them "**Z**-equations" for short:

DEFINITION: A **Z**-*equation* is an equality of polynomials with integer coefficients.

Note that if you have an equality of polynomials with rational number coefficients, you can transform it into an equivalent **Z**-equation by multiplying both sides through by a common denominator. For example, the equation $\frac{y^2}{3} = \frac{x^3}{3} - x + \frac{14}{3}$ can be transformed into (6.1) by multiplying both sides by 3.

One of the main problems in number theory is finding and understanding all solutions of **Z**-equations. A solution to an equation can be exemplified using (6.1). We say that $x = 2$, $y = 4$ is a solution because if we replace x by 2 and y by 4 on both sides of the equation and do the arithmetic, we get $16 = 16$, which is true. We can talk about the "solution set S" to a given **Z**-equation. This is the set whose elements are all the solutions of that equation.

Wait a minute. Here's another solution to equation (6.1): $x = 1$, $y = \sqrt{12}$. It is a solution, but the y-value is not an integer, and not even a rational number. Is it in the solution set S?

To quote Humpty Dumpty, "It all depends on which is to be master." *We* can decide what values we will allow the unknowns to take. If we decide that S will be the set of real-number solutions, then we will allow $x = 1, y = \sqrt{12}$ into S, but if we want S to contain only rational-number or integral solutions, we will not.

We need a good notation to clarify in any given context which solutions are allowed. We do this by symbolizing the sets of possible kinds of numbers that we are interested in. We let **Z** be the set of integers, **Q** the set of rational numbers, **R** the set of real numbers, and **C** the set of complex numbers. There are other number systems we will consider, too, especially $\mathbf{F}_p$, the set of integers modulo p, for various primes p.

Varieties

Suppose we have fixed our attention on a particular **Z**-equation. We write $S(\mathbf{Z})$ for the set of all integral solutions of that equation, $S(\mathbf{Q})$ for the set of all rational solutions of it, and so on. We call S an "algebraic variety," because for the various choices of number systems, we get various sets of solutions. We say that the given equation defines the variety S.

> **DEFINITION**: The *variety* S defined by a **Z**-equation (or a system of **Z**-equations) is the function that assigns to any number system A the set of solutions $S(A)$ of the equation (or system of equations).

For example, take the equation $x^2 + y^2 = 1$. We know that the set $S(\mathbf{Z})$ cannot be very big, because $x^2 \geq 0$ and $y^2 \geq 0$. Trial-and-error with small numbers tells us the whole story: $S(\mathbf{Z})$ contains four elements, and it is $\{(1, 0), (-1, 0), (0, 1), (0, -1)\}$, where the pair (a, b) stands for the solution $x = a$ and $y = b$.

What about $S(\mathbf{Q})$? The answer can be found in Euclid, though not written quite like this. It can be shown that

$$S(\mathbf{Q}) = \left\{ \left(\frac{1-t^2}{1+t^2}, \frac{2t}{1+t^2} \right) : t \text{ any rational number} \right\} \bigcup \{(-1, 0)\}.$$

The symbol $\bigcup$ denotes the *union* of the two sets, that is, the set that contains all elements of both sets.

How about $S(\mathbf{R})$? This is easier: If x is any number between -1 and 1, we can let $y = \sqrt{1-x^2}$ or $y = -\sqrt{1-x^2}$. (By convention, if a is positive, then $\sqrt{a}$ always stands for the positive square root of a.) We will in the future abbreviate the preceding two solutions by writing $y = \pm\sqrt{1-x^2}$. Incidentally, there is a fancier way to write this same solution set using trigonometry:

$$S(\mathbf{R}) = \{(\cos\theta, \sin\theta) : 0 \leq \theta < 2\pi\}.$$

Another example is $S(\mathbf{C})$. Here, we can let x be any complex number at all, and let $y = \pm\sqrt{1-x^2}$. The square roots will always exist, because we are allowing complex numbers.

What about modular arithmetic? If we want to compute $S(\mathbf{F}_2)$, for example, we can let x and y run through all of the elements of $\mathbf{F}_2$ (which does not take very long, because there are only the two elements 0 and 1 in $\mathbf{F}_2$), and compute x^2+y^2, and see if we get 1. We find out that $S(\mathbf{F}_2) = \{(1,0),(0,1)\}$. Similarly, we can compute that $S(\mathbf{F}_3) = \{(1,0),(2,0),(0,1),(0,2)\}$.

EXERCISE: List all of the elements of $S(\mathbf{F}_5)$.

SOLUTION: This is just trial-and-error. After running through all of the possibilities, we get $S(\mathbf{F}_5) = \{(1,0),(4,0),(0,1),(0,4)\}$.

EXERCISE: List all of the elements of $S(\mathbf{F}_7)$.

SOLUTION: This is a bit more interesting, and again trial and error gives all the solutions. We get $S(\mathbf{F}_7) = \{(1,0), (6,0), (0,1), (0,6), (2,2), (2,5), (5,2), (5,5)\}$.

We can state Fermat's Last Theorem in the language of varieties. Fermat's Last Theorem is the following assertion:

For any positive integer n, let the variety V_n be defined by

$$x^n + y^n = z^n.$$

Then if $n > 2$, $V_n(\mathbf{Z})$ contains only solutions where one or more of the variables is 0.

One reason to state Fermat's Last Theorem in this way is that it is easier to study $V_n(\mathbf{R})$, $V_n(\mathbf{C})$, or $V_n(\mathbf{F}_p)$ than $V_n(\mathbf{Z})$. Then some information about $V_n(\mathbf{Z})$ (the solution set we are really interested in) can be derived from the preceding three sets, using advanced theorems in number theory and algebraic geometry. The complete proof of Fermat's Last Theorem, however, required Galois representations and many other additional ideas, as we shall see.

Systems of Equations

Next, we want to consider *systems* of equations. In English, we often connect our assertions with "and": *The first president of the United States was George Washington and the sixteenth president was Abraham Lincoln*. We are asserting that both statements are true. We can do this with unknowns as well. Consider:

1. X = the first president of the United States.
2. Y = the sixteenth president of the United States.

This system has the solution X= George Washington, Y=Abraham Lincoln. Solving the system means solving both equations at once. In this example, the equations do not share any unknowns, so solving both is the same as solving each one separately. We say that the equations are *uncoupled*. That is not very interesting. Consider now:

1. X = the first president of the United States.
2. X = the sixteenth president of the United States.

Now the equations are *coupled*, and there is no solution to the system, although each equation can be solved separately. (By the way, there is a pair of numbers that can be put in place of "first" and "sixteenth" for which the system has the unique solution X = Grover Cleveland.) Finally, consider:

1. X = the first president of the United States.
2. X = the commanding officer of the colonists during the American Revolution.

This coupled system does have a solution, again unique, namely, X = George Washington.

Of course, solution sets to systems of equations need not contain only one element. To put it another way, solutions need not be *unique*. For a given system of equations, there may be no solutions, exactly one, more than one but still finitely many, or infinitely many solutions. For example, consider the system of **Z**-equations:

1. $x + y = z$.
2. $x^2 + 2xy + y^2 = z^2$.

Any solution to the first equation is also a solution to second equation. Because any three numbers for which the sum of the first two is the third gives a solution to the first equation (hence to both), we see there are infinitely many solutions to the two equations, even if we allow only integer solutions.

Here are some more examples of systems of equations and the varieties they define. First, a nonexample:

1. $x^2 + y^2 = 1$.
2. $x > 0$.

This system makes perfect sense, but we do not allow it because it involves an inequality in the second statement. Inequalities have no meaning when applied to some number systems, for instance **C** or $\mathbf{F}_p$. So this system does not define an algebraic variety.

Next, an example with more than two equations in the system:

1. $x^2 + y^2 + z^2 = w$.
2. $w^4 = 1$.
3. $x + y = z$.

Just for fun, let us find $S(\mathbf{Z})$ for this system. From (2) we see that $w = 1$ or $w = -1$. There are no solutions to (1) if $w = -1$, and if $w = 1$, we get that x, y and z are all 0 or ± 1, and exactly two of them equal 0. But then (3) is impossible. So $S(\mathbf{Z})$ is the empty set.

Now, we find $S(\mathbf{R})$ for this system. Again, from (2) we get that $w = 1$ or $w = -1$, and again equation (1) has no solutions when $w = -1$. We now substitute $x + y$ for z in (1), which we are allowed to do because of (3). Thus, $S(\mathbf{R})$ is the same as the set of **R**-solutions of the following single equation:

4. $x^2 + y^2 + (x + y)^2 = 1$.

or equivalently

5. $x^2 + y^2 + xy = 1/2$.

The solutions to (5) may be graphed on (x, y)-graph paper as an ellipse. There are infinitely many **R**-solutions to (5), and so $S(\mathbf{R})$ is infinite.

There are even more **C**-solutions. First, from (2) we see that w can be any one of the four fourth roots of 1.[3] The same reasoning as before tells us that for each choice of a fourth root of 1 to be the value of w, we will get an infinite number of possibilities for (x, y, z).

It is easy to write down systems or even single equations whose solution set is very hard to determine, or even unknown.

EXERCISE: Let S be the variety defined by

1. $x^{17} + 33x^2y^3 - xyz = 44$.
2. $x^3 + y^3 = z^2 + 137$.

What is $S(\mathbf{Z})$?

SOLUTION: We have no idea—it is probably very hard to figure this out. Even if it turns out that by some trick you can solve this system of equations (maybe reducing modulo a prime could show there are no integer solutions), just replace it by an even more complicated system of your invention.

In summary, given a system of **Z**-equations, we can consider the variety S of simultaneous solutions. Then for any number system A, we get the set $S(A)$ of simultaneous solutions where the values of the variables are drawn from the system A. We will see many more examples of varieties as we go along. We will try to get a feel for some varieties that are particularly interesting because of their structure.

Equivalent Descriptions of the Same Variety

Many different systems of equations can define the same variety S. It is important to understand this concept, so we give a few examples. For instance, any set of inconsistent equations will define the empty variety.

[3]The 4 fourth roots of 1 are $1, -1, i$, and $-i$, as you can see by raising each of them to the fourth power. It can be proven that every nonzero complex number has n different nth complex roots for all $n \geq 1$.

EXAMPLE: If S is the variety defined by the system:

1. $x = y$
2. $x = y + 1$

then $S(A)$ is the empty set for any number system A. If instead we consider the system

1. $x = y^2$
2. $x = y^2 - 1$

and let T be the variety it defines, then T is also the empty variety. So $S = T$ even though the systems of equations are different.

EXAMPLE: Consider the two systems

1. $2x + 3y = 7.$
2. $x - y = 6.$

and

1′. $3x + 2y = 13.$
2′. $5x + 5y = 20.$

We get (1′) by adding (1) and (2). We get (2′) by doubling (1) and adding it to (2). Similarly, we can get the first system of equations from the second by

- subtracting (1′) from (2′) to get (1);
- doubling (1′) and subtracting (2′) to get (2).

Therefore, the two systems are totally equivalent: they have the same solution sets in any number system. They define the same variety.

EXAMPLE: Yet another:

1. $x^2 + y^3 = 0.$
1′. $(x^2 + y^3)^2 = 0.$

Because only 0^2 is 0, these two equations define the same variety, whatever it is.

Consider the equation $x^3 + y^3 = 0$. It defines a variety, which we can call V. We can define a new variable $z = x + y$. For every possible pair of values for x, y we have the corresponding pair of values for z, y. For example, $x = 2$, $y = 5$ corresponds to $z = 7$, $y = 5$. This defines a "transformation" from the variables (x,y) to the variables (z,y). This transformation is a one-to-one correspondence between sets of possible values of the variables. Why? Clearly (x,y) corresponds to just one possible (z,y), namely, $(x+y,y)$. On the other hand, given (z,y) it corresponds to just one possible (x,y), namely, $(z-y,y)$. For instance, $z = 99, y = 77$ corresponds to $x = 22$, $y = 77$. We call such a variable transformation as this an *invertible transformation*.

We can then transform the equation $x^3 + y^3 = 0$ to an equivalent equation in the new variables, using the rule: $x = z - y$. The new equation is $(z-y)^3 + y^3 = 0$. (If desired, we can multiply out the first cube and collect terms to get the logically equivalent equation $z^3 - 3z^2y + 3zy^2 = 0$.) Because the variable transformation is invertible,[4] there is a one-to-one correspondence of solutions to this equation in any number system A with solutions to the original equation in A. In other words, this transformed equation defines an equivalent variety. We view these as *isomorphic varieties*, or, more colloquially, as "the same" varieties. The equations and solution sets are not equal—indeed, the variables in one case are x and y and in the other case z and y. We say the two varieties are "isomorphic" (which etymologically means they have the "same shape"); the transformation guarantees that you can go back and forth in a one-to-one correspondence from one solution set to the other, in any number system.[5]

If we have the information that two systems define isomorphic varieties, and one system is much easier to solve than the other, we have discovered something significant about the more difficult looking system.

---

[4]This means reversible.

[5]Our examples with the empty variety show that two systems of equations can define isomorphic varieties even if they are not connected by an invertible variable transformation.

Finding Roots of Polynomials

The easiest *general* class of varieties to look at would be those defined by a single **Z**-equation in a single variable, for instance,

$$x^3 + x - 2 = 0. \tag{6.2}$$

The study of this type of variety is dominated by the concept of the *Galois group* (see chapters 8 and 13).

We now review the following terminology:

DEFINITION: If $f(x)$ is a polynomial, the *roots* of $f(x)$ are those numbers c so that $f(c) = 0$.

The roots of $f(x)$ are the solutions to the polynomial equation $f(x) = 0$. Now, there is the particularly simple equation $x^n = a$, and we call a solution to it an nth root of a. Sometimes "root" means a square root or cube root, and so on, and sometimes it means a root of a more general polynomial. You should be able to tell which root is meant from the context.

It is helpful to symbolize the polynomial we are studying by a single letter, say p. If we want to remember the name of the variable, we can write $p(x)$. For instance, $p(x)$ might denote the polynomial $x^3 + x - 2$. Then equation (6.2) can be written $p(x) = 0$.

This looks like functional notation, and it is. If x is a variable, $p(x)$ just stands for the polynomial p, but if a is a number, $p(a)$ stands for the number you get by substituting a for x in p. For instance, if $p(x) = x^3 + x - 2$ then $p(0) = 0^3 + 0 - 2 = -2$, $p(1) = 1^3 + 1 - 2 = 0$, and in general, $p(a) = a^3 + a - 2$.

For example, consider the variety of solutions, call it S, to the equation $p(x) = 0$, where $p(x)$ is the polynomial discussed in the preceding paragraph. That is, if A is any number system, describe the set $S(A)$, which is the set of all elements a of A such that $p(a) = 0$.

We can use simple algebra and some guessing to find $S(\mathbf{Z})$: We have just seen that $p(1) = 0$. So $S(A)$ contains the number 1. High-school algebra now tells us that we can divide $p(x)$ by $x - 1$ and we are guaranteed that it will go in without remainder. Doing that we get the quotient $x^2 + x + 2$.

In other words, $p(x) = x^3 + x - 2 = (x-1)(x^2+x+2)$. Now for any integer a, $p(a) = 0$ if and only if $(a-1)(a^2+a+2) = 0$, because $p(a) = (a-1)(a^2+a+2)$. A product of two integers is 0 only if one or both of them is 0. (This is true in any number system that we will use in this book.) So if we try to solve $(a-1)(a^2+a+2) = 0$, we see that either $a = 1$ (which we already knew was a possibility) or else $a^2 + a + 2 = 0$.

The quadratic formula[6] will tell us what all the solutions of the quadratic equation are: $a = \frac{-1 \pm \sqrt{1-8}}{2}$. But the negative number -7 has no real square root, let alone an integer square root. So this option leads us nowhere in finding more integer solutions to $p(x) = 0$.

But we have been exhaustive: If there were any other solutions, we would have found them because $p(x) = 0$ if and only if $x - 1 = 0$ or $x^2 + x + 2 = 0$. In conclusion, we can confidently say that $S(\mathbf{Z}) = \{1\}$. In fact, for any number system A,[7] this same method will find $S(A)$ for us, and you can see that $S(A)$ will have at least one element (since every number system has to have the number 1 in it) and may possibly have two or three elements, but no more.

EXERCISE: Find a number system A for which $S(A)$ has exactly two elements.

SOLUTION: Use trial-and-error on various $\mathbf{F}_p$'s. One answer is $\mathbf{F}_2$: $S(\mathbf{F}_2) = \{0, 1\}$. Another is $\mathbf{F}_7$: $S(\mathbf{F}_7) = \{1, 3\}$.

Are There General Methods for Finding Solutions to Systems of Polynomial Equations?

We were very lucky in our attempt to solve (6.2) because

1. we could guess one of the solutions;

---

[6]The quadratic formula states that the solutions of the equation $ax^2 + bx + c = 0$ are $x = \frac{-b \pm \sqrt{b^2 - 4ac}}{2a}$.

[7]As long as we are allowed to divide by 2 in A; otherwise, the quadratic formula does not work.

2. we knew the quadratic formula;
3. the original equation was only degree 3 to start with (i.e., the highest power of x appearing was x^3).

Are there general methods to solve systems of polynomial equations?

Answer: No, no, no! That difficulty is what makes this whole subject so interesting. With some exaggeration, you can say that this difficulty is the reason that whole chunks of mathematics such as algebraic number theory and algebraic geometry have come into being.

How do we know there is no general method of solving an arbitrary system of polynomial equations? It all depends on what is meant by "general."

First, we have to distinguish between exact solutions and approximate solutions. Suppose that $p(x)$ is a polynomial. Especially with the advent of computers, there are ways of finding numbers a so that $p(a)$ is approximately 0. These ways may be general, meaning: You give the computer the polynomial $p(x)$ and you give it a positive number, a "tolerance," as small as you like, say 0.001. Then the computer will grind away and perhaps find numbers a so that $p(a)$ is between -0.001 and $+0.001$. There are two problems with this from our more "exacting" point of view.

1. The computer may never report back, even if there is a solution. Suppose we give it a gigantic $p(x)$, of degree 100,000,000, with all kinds of large integer coefficients, and suppose we make our tolerance very small, say 0.00000000000000001. Then perhaps the computer will not find a solution within our lifetimes.
2. Even if the computer reports back, it is only giving us an approximate solution.[8] That does not have to bother us if we are building a bridge or doing our taxes—approximations are fine if we have set the tolerances properly. But the game of number theory is a game where we want to know about exact solutions.

[8]We are assuming that the computer is using "floating point arithmetic."

Second, what is “general?” Here is a “general” polynomial equation solver: There is a woman named Sybil, who lives down the street from us. We do not know where she gets her amazing power, but this is what she can do: We write down any system of **Z**-equations and show them to her. We also tell her what number system A we are interested in. To make this example simple, assume we are only interested in $A = \mathbf{Z}$, that is, integral solutions. Ten seconds later, she writes down a list of integer solutions and tells us these are all the elements of the solution set $S(\mathbf{Z})$. We check: Her integers all satisfy the equations we have given her to solve. And we have never found an integral solution not on her list.

If we give Sybil the gigantic equation of degree 100,000,000 that we gave the computer, she still only waits 10 seconds, but then sometimes she writes and writes and writes, because there are so many digits in the answer. Maybe the answer is so big, it will take her a billion billion years to write it down. We do not have that long to wait. But we can hardly blame her for that—she “knows” the answer and it just takes a while to record it. And maybe the system of equations that we give Sybil has *infinitely* many solutions, and then she will be writing forever!

As far as we know, Sybil is a general method for solving all systems of **Z**-equations in one or several variables. But we cannot be sure, because we do not know how she does it, and we cannot be positive she will always be right. And even when she gives us a listing of $S(\mathbf{Z})$, we cannot be *sure* she has not left a solution out, even though *we* cannot find one. (Of course, we can check her for simple systems, such as one polynomial equation of degree 2, but we are talking “general” here.)

So even if Sybil existed, we still could not be sure she was a general solver. But a true oracle might exist, perhaps in the next galaxy, so we cannot say “no, no, no” with such certainty unless we restrict what we mean by a “general method.”

Nowadays, there is an agreed-upon answer: A *general method* means something that we can program a computer to do. There is a surprising theorem that, using this definition of a general method, there cannot be a general method of finding all integer solutions to all systems of polynomial **Z**-equations in many variables. (Of course, for polynomial **Z**-equations in *one* variable, we can bound the size of the roots and then search a finite domain, so if there is only one variable, then there is a general method for finding $S(\mathbf{Z})$.)

On a purely number-theoretical level, leaving philosophy and logic behind, we also have the famous theorem of Abel and Ruffini: Unlike quadratic polynomials, for which we can use the quadratic formula, for polynomials $f(x)$ of degree 5 or greater, there is no formula involving just addition, subtraction, multiplication, division, and nth roots $(n = 2, 3, 4, \ldots)$ that can solve $f(x) = 0$ in general.

Deeper Understanding Is Desirable

Finally, even if we had a Sybil, what good would it do us from a theoretical point of view? We would like to understand *why* these solutions exist, if they do, or why they do not, if they do not, for particular classes of equations. And we would like to understand the inner structure of the solution sets, and of the variety. If we know enough about these things, we can then prove beautiful theorems and understand *why* they are true. This is what mathematics is all about. If we had asked Sybil if Fermat's Last Theorem were true before Wiles proved it in 1995, and she said "Yes," would we be satisfied? No. Although we might stop trying to prove it false, we would continue to try to prove it true, in order to understand *why* it is true.

This is the true role of computer-aided proofs, such as the proof of the Four Color Theorem.[9] They are like Sibyl. It could be that when we are convinced, by Sybil or by computer, that a given theorem is true, we might lose interest in the theorem altogether. On the other hand, mathematicians are always coming up with new proofs of old theorems, when those theorems remain interesting and the new proofs enhance our understanding of them.

[9]The Four Color Theorem states that it is always possible to color any map with four colors in such a way that no two neighboring countries will have the same color. The proof uses a computer to check that many types of maps can be colored in this way.

To go back to a single **Z**-equation of degree d in a single variable, defining a variety S: If A is any field, we can prove that $S(A)$ can never have more than d solutions in it. (Basically we just keep dividing by $x - a$ for various solutions a.) It is quite interesting to study $S(A)$ simply as a set. What structure does a finite set have? Only one thing: How many elements are in it. But knowing how many elements are in $S(A)$ tells us if there are none, exactly one, or more than one solution. These can be burning questions. The security of the credit card number that you just entered on your computer when you made a purchase over the Internet could conceivably depend on the answer to such a question.

The amazing discovery of Galois is that there is more structure to $S(A)$. As we shall see, $S(A)$ is not just a set; it is the basis for defining a representation of a certain group, called the *Galois group*. Even if we only wanted to know how many elements $S(A)$ has, in many cases this is a difficult question that can only be approached (as far as anyone today knows) via the Galois group and its representations. And the same thing is true about **Z**-varieties of all kinds. That finally leads us to the central subject of this book: Galois groups and their representations. But before we go on to Galois groups, we want to look at some very simple, very interesting, and very important **Z**-varieties in the next chapter. Then, after we explain Galois groups, we will look at another series of very interesting and very important, though not so very simple, **Z**-varieties: *elliptic curves*. These two kinds of varieties will give us some of our main examples to help us understand Galois groups and their representations.

Chapter 7

QUADRATIC RECIPROCITY

Road Map

We begin to explore **Z**-equations by looking very closely at a fairly simple example: $x^2 = a$, where a is some constant integer. It is hard to get simpler than this equation and still have something interesting. In this case, the interest lies not in solving the equation in **Z**, **R**, or **C**, all of which are easy, but in $\mathbf{F}_p$ for various prime numbers p.

Quadratic reciprocity refers to a mysterious relationship between the solutions to $x^2 = q$ in $\mathbf{F}_p$ and $x^2 = p$ in $\mathbf{F}_q$, where p and q are two different odd prime numbers. Quadratic reciprocity was the first of all reciprocity laws, and it is closely connected with the theory of Galois representations. We still have not defined these things, but we want to have the example of quadratic reciprocity under our belt before we do.

Quadratic reciprocity is part of "classical" number theory going back to the eighteenth century. It inspired the many other reciprocity laws that eventually developed into a large part of modern number theory.

The Simplest Polynomial Equations

Let us discuss the solutions of polynomial equations. Since the time of the French mathematician and philosopher René Descartes, mathematicians have realized that the *degree* of a polynomial is a good indication of how complicated it is. Recall that the degree of a polynomial $f(x)$ of one variable is the highest power of x that appears. For example, $10x^{33} - 100x^7 + 1{,}729$ has degree 33.

A polynomial of degree 0 is simply a constant. Because a rose is a rose is a rose, there is not much to say about a single constant, and we can go on to the next case.

A polynomial of degree 1 is, for example, $5x + 3$. The set of solutions is again completely understandable. In any field at all (excluding those where we are not allowed to divide by 5, such as $\mathbf{F}_5$), this polynomial has exactly one root, which we can find by dividing -3 by 5.

So the first nontrivial polynomials are those of degree 2. We will take this chapter to study just one aspect of quadratic equations. As you will see, these matters are truly not trivial, and lead to some very pretty and historically crucial bits of mathematics. In fact, the subject matter here, quadratic reciprocity, is the tip of the gigantic iceberg of "reciprocity laws" that we will explain later.

Rather than taking the most complicated degree 2 equation, which would look like $Ax^2 + Bx + C = 0$, choose just one integer, call it a, and look at the equation $x^2 - a$. If a is 0, then the solution set of $x^2 - 0 = 0$ will just be $x = 0$, and so it will contain one number. Otherwise, the variety defined by $x^2 - a$ will usually contain either two numbers or none, depending on whether or not a is a square.

For example, 1 is 1^2 in every number system. If we take $a = 1$, and look at $x^2 - 1$, then $S(\mathbf{Z}) = \{1, -1\}$, and $S(\mathbf{Q})$, $S(\mathbf{R})$, and $S(\mathbf{C})$ will be the same set. If we take $x^2 + 1$, for instance (meaning that $a = -1$; we never said that a had to be positive), then $S(\mathbf{Z})$ has no elements, and $S(\mathbf{Q})$ has no elements, and $S(\mathbf{R})$ has no elements, and $S(\mathbf{C}) = \{i, -i\}$. What happened here is that -1 is a square in $\mathbf{C}$ but not in $\mathbf{Z}$, $\mathbf{Q}$, or $\mathbf{R}$.[1]

We are going to play this game now using $\mathbf{F}_p$ for various primes p, after fixing some integer a. It is not so interesting to let $p = 2$: if $a = 0$, then $S(\mathbf{F}_2) = \{0\}$, and if $a = 1$, then $S(\mathbf{F}_2) = \{1\}$. Therefore,

---

[1]In general, if R is a field and $a = b^2$ with a and b elements of R, then $x^2 - a = (x - b)(x + b)$ equals 0 exactly if $x = b$ or $x = -b$, and then $S(R) = \{b, -b\}$. There will be two different elements in $S(R)$ unless $b = -b$, which only happens if R has characteristic 2 or if $b = 0$. If a is not a square in R, then $x^2 - a = 0$ has no solutions and $S(R)$ is empty.

we can restrict our attention to odd primes. For the remainder of this chapter, assume that p is an odd prime.

When Is −1 a Square mod p?

We return to x^2+1, and try various values of p. By trial-and-error, we find that $S(\mathbf{F}_3)=\varnothing$, and $S(\mathbf{F}_5)=\{2,3\}$, and $S(\mathbf{F}_7)=\varnothing$, and $S(\mathbf{F}_{11})=\varnothing$, and $S(\mathbf{F}_{13})=\{5,8\}$. Here, $\varnothing$ is the standard symbol for the *empty set*. In other words, "$S(\mathbf{F}_{11})=\varnothing$" means that there is no element in $\mathbf{F}_{11}$ whose square equals -1. Or, equivalently, whatever integer b you take, you will never find that b^2+1 is evenly divisible by 11.

EXERCISE: Why does $S(\mathbf{F}_3)=\varnothing$, and $S(\mathbf{F}_5)=\{2,3\}$?

SOLUTION: To compute $S(\mathbf{F}_3)$, we substitute the three elements of $\mathbf{F}_3$ into the polynomial x^2+1, and see if we ever get 0 modulo 3:

$$0: 0^2+1=1\not\equiv 0 \pmod 3$$

$$1: 1^2+1=2\not\equiv 0 \pmod 3$$

$$2: 2^2+1=5\not\equiv 0 \pmod 3$$

We have now tried out everything in $\mathbf{F}_3$, and nothing worked, so $S(\mathbf{F}_3)=\varnothing$.

What about $S(\mathbf{F}_5)$? We try the five elements of $\mathbf{F}_5$ and see what happens:

$$0: 0^2+1=\ \ 1\not\equiv 0 \pmod 5$$

$$1: 1^2+1=\ \ 2\not\equiv 0 \pmod 5$$

$$2: 2^2+1=\ \ 5\equiv 0 \pmod 5$$

$$3: 3^2+1=10\equiv 0 \pmod 5$$

$$4: 4^2+1=17\not\equiv 0 \pmod 5$$

So that is how to see that $S(\mathbf{F}_5)=\{2,3\}$.

We can observe that if $S(\mathbf{F}_p) \neq \varnothing$, then the two elements in $S(\mathbf{F}_p)$ add up to p.

EXERCISE: Suppose that $S(\mathbf{F}_p) = \{n, m\}$. Show that $n + m = p$.

SOLUTION: Suppose that we know that $S(\mathbf{F}_p)$ contains the number n. This means that $n^2 + 1 \equiv 0 \pmod{p}$. Now, we plug the number $p - n$ into the same polynomial $x^2 + 1$. We get $(p - n)^2 + 1 \equiv (-n)^2 + 1 = n^2 + 1 \equiv 0 \pmod{p}$. This means that $p - n$ is an element of $S(\mathbf{F}_p)$. So the two numbers in $S(\mathbf{F}_p)$ are n and $p - n$, and those two numbers add up to p. Because p is odd, we know that n and $p - n$ must be unequal numbers (one is odd and the other is even), so we really have two different solutions.

One of the truly inspired notions in mathematics was to step back, and not worry about the *particular* numbers contained in $S(\mathbf{F}_p)$. Instead, we can ask: For which primes p is $S(\mathbf{F}_p) = \varnothing$, and for which is it not?

EXERCISE: List all of the primes under 100, and for each odd prime on your list, decide if $S(\mathbf{F}_p) = \varnothing$ or not. Use trial-and-error—and a computer, or at least a calculator, for the larger primes.

SOLUTION: You will find that $S(\mathbf{F}_p) = \varnothing$ if $p = 3$, 7, 11, 19, 23, 31, 43, 47, 59, 67, 71, 79, and 83, and $S(\mathbf{F}_p) \neq \varnothing$ if $p = 5$, 13, 17, 29, 37, 41, 53, 61, 73, 89, and 97.

The amazing thing here is that there is a pattern in these two lists! The first list consists of those primes p so that $p \equiv 3 \pmod{4}$, and the second list is those primes p so that $p \equiv 1 \pmod{4}$. We will not prove this observation in this book, but we will assume it from now on. The proof is somewhat complicated. The theorem that we are aiming at, called "quadratic reciprocity," is much deeper, and its proof is correspondingly more complicated—we will not be giving that proof either.

The Legendre Symbol

There is a clever bit of notation, going back to the French mathematician Adrien-Marie Legendre (a contemporary of Gauss's), that lets us summarize what we just discovered. We will also see that it does more than summarize; it has an important multiplicative property (7.1) that will be interpreted in chapter 19. It is called the *Legendre symbol* in his honor. Remember that we noticed that after we select some integer a, the variety $S(\mathbf{F}_p)$ corresponding to the polynomial $x^2 - a$ can have 0, 1, or 2 elements. Here is how Legendre chose to record this fact. He defined the Legendre symbol $\left(\frac{a}{p}\right)$ according to the following formula:[2]

$$\left(\frac{a}{p}\right) = \begin{cases} -1 & \text{if the number of solutions to } x^2 - a \equiv 0 \pmod{p} \text{ is } 0. \\ 0 & \text{if the number of solutions to } x^2 - a \equiv 0 \pmod{p} \text{ is } 1. \\ 1 & \text{if the number of solutions to } x^2 - a \equiv 0 \pmod{p} \text{ is } 2. \end{cases}$$

Thus, what we were exploring in the previous section was $\left(\frac{-1}{p}\right)$.

> **WARNING**: The Legendre symbol $\left(\frac{a}{p}\right)$ is *not a* divided by p. Legendre used a fraction symbol but put in the parentheses to remind you that it is *not* division but—the Legendre symbol![3]

---

[2]See our previous footnote to see why $S(\mathbf{F}_p)$ cannot have more than two elements. Also note the following facts:

- Because p is odd, $b \not\equiv p - b \pmod{p}$ unless $b \equiv 0 \pmod{p}$.
- $p - b \equiv -b \pmod{p}$, i.e., $p - b = -b$ as elements of the field $\mathbf{F}_p$.

So if a is a nonzero square in $\mathbf{F}_p$, then it has two unequal square roots b and $-b$ for some b in $\mathbf{F}_p$.

[3]If you leave out the horizontal bar, you get the symbol $\binom{a}{p}$, which is reserved for something completely different called a *binomial coefficient*.

In other words,

- $\left(\dfrac{a}{p}\right) = -1$ if $S(\mathbf{F}_p)$ contains no elements.
- $\left(\dfrac{a}{p}\right) = 0$ if $S(\mathbf{F}_p)$ contains one element. This happens only if a is a multiple of p.
- $\left(\dfrac{a}{p}\right) = 1$ if $S(\mathbf{F}_p)$ contains two elements.
- Because $\left(\dfrac{a}{p}\right)$ depends only on a modulo p, $\left(\dfrac{a}{p}\right) = \left(\dfrac{a+kp}{p}\right)$ for any integer k.[4]

You might wonder why Legendre chose to shift by 1, rather than just letting $\left(\dfrac{a}{p}\right)$ be the number of elements in $S(\mathbf{F}_p)$. One answer is the following easy pair of equations:

$$\left(\frac{0}{p}\right) = 0, \qquad \left(\frac{1}{p}\right) = 1.$$

But there is more to it than that.

Digression: Notation Guides Thinking

There are some very famous instances of how proper choices of notation can help speed the progress of mathematics. Someone outside of the field might not even realize that there is a distinction between the notations that mathematicians use and the subject itself.

For example, algebraic notation—x, y, and all the rest—is notation, not mathematics. Greek mathematicians did not have that symbolism, and they were able to express problems in words and solve them without using symbols. Today, of course, the use of x to stand for an unknown thing is so widespread that it has passed into popular culture. Using letters to stand for quantities makes it

[4]Pay careful attention to this fact; we will be using it repeatedly, both in proofs and computations.

possible to do a lot of mathematics that would otherwise be tedious or too complicated to comprehend when stated in words.

Another example is powers. We write x^2 to mean x times x, without giving it a second thought. But as we write this, we can think of the exponent as another variable, and write x^n. We can write down the *power law of exponents*: $x^n x^m = x^{n+m}$. Influenced by this symbolism, we can then start thinking about x^n where n is no longer an integer. We can write down $x^{1/2}$ and give a meaning to that symbol: $x^{1/2}$ means the square root of x.

A fascinating example of the advantages and disadvantages of notation occurred when calculus was invented (or discovered). Newton used one symbol to stand for the process of *differentiation*; Leibniz used a different symbol. Newton was perhaps the greater mathematician and scientist, but his notation was quite hard to use. Leibniz's notation was so good that every year, thousands of students compute derivatives without ever fully understanding them, because the notation almost forces the students to do the right computation—at least most of the time!

We will see immediately that Legendre's notation cleverly helps us to notice and remember a certain multiplicative structure more easily than we would have with the naïve "unshifted" notation.

Multiplicativity of the Legendre Symbol

Legendre's notation helps clarify what happens when p is fixed and the number a varies. In particular, after playing around with some examples, you may discover—and it is possible to prove—that

$$\left(\frac{a}{p}\right)\left(\frac{b}{p}\right) = \left(\frac{ab}{p}\right). \tag{7.1}$$

EXERCISE: Show that $\left(\frac{2}{7}\right)\left(\frac{5}{7}\right) = \left(\frac{10}{7}\right)$.

SOLUTION: You can check that $3^2 - 2 \equiv 0 \pmod 7$ and $4^2 - 2 \equiv 0 \pmod 7$. Therefore, $\left(\frac{2}{7}\right) = 1$. If you try the seven numbers from 0 to 6, you will find that $x^2 - 5 \equiv 0 \pmod 7$ has no solutions, and therefore $\left(\frac{5}{7}\right) = -1$. And, finally, if you try all seven possibilities in the congruence $x^2 - 10 \equiv 0 \pmod 7$, again you will find that there are no solutions, so $\left(\frac{10}{7}\right) = -1$. And that checks (7.1): $1 \cdot -1 = -1$.

We now return to our polynomial $x^2 + 1$, which corresponds to choosing $a = -1$. In terms of Legendre's notation, we have discovered that

$$\left(\frac{-1}{p}\right) = \begin{cases} 1 & \text{if } p \equiv 1 \pmod 4 \\ -1 & \text{if } p \equiv 3 \pmod 4. \end{cases}$$

Before we continue, remember that if a is not divisible by p, then $\left(\frac{a}{p}\right)$ is not 0. In particular, $\left(\frac{-1}{p}\right)$ is never 0.

When Is 2 a Square mod p?

The next number to consider is 2. We present you with the following challenge if you like computer programming—otherwise, just read our solution:

EXERCISE: List all of the odd primes under 100, and see for which p you have $\left(\frac{2}{p}\right) = -1$, and for which p you have $\left(\frac{2}{p}\right) = 1$.

SOLUTION: It turns out that $\left(\frac{2}{p}\right) = 1$ if $p = 7, 17, 23, 31, 41, 47, 71, 73, 79, 89$, and 97. On the other hand, $\left(\frac{2}{p}\right) = -1$ if $p = 3, 5, 11, 13, 19, 29, 37, 43, 53, 59, 61, 67$, and 83.

You could stare at this list for a long time before discovering the pattern, even when you know there *is* a pattern. If you did not know there was something going on, you would have an even harder time seeing the pattern. The answer is:

$$\left(\frac{2}{p}\right) = \begin{cases} 1 & \text{if } p \equiv 1 \pmod 8 \\ -1 & \text{if } p \equiv 3 \pmod 8 \\ -1 & \text{if } p \equiv 5 \pmod 8 \\ 1 & \text{if } p \equiv 7 \pmod 8. \end{cases}$$

What this equation means is that if you have an odd prime p, and you want to know whether or not 2 has two square roots in $\mathbf{F}_p$, then you see what p is congruent to modulo 8, and look at the corresponding line on the right-hand side of the equation to get the value of the Legendre symbol $\left(\frac{2}{p}\right)$. In other words, $\left(\frac{2}{p}\right) = 1$ if $p \equiv 1$ or $7 \pmod 8$ and $\left(\frac{2}{p}\right) = -1$ if $p \equiv 3$ or $5 \pmod 8$. For example, $47 \equiv 7 \pmod 8$, and sure enough, $7^2 \equiv 2 \pmod{47}$ and $40^2 \equiv 2 \pmod{47}$.

When Is 3 a Square mod *p*?

We now move on even further to $a = 3$, and try again:

EXERCISE: Use the list of odd primes under 100 and decide for which of those primes the equation $\left(\frac{3}{p}\right) = 1$ holds, and for which you have $\left(\frac{3}{p}\right) = -1$.

SOLUTION: This time, our computer tells us that $\left(\frac{3}{p}\right) = 1$ when $p = 11, 13, 23, 37, 47, 59, 61, 71, 73, 83$, and 97, and $\left(\frac{3}{p}\right) = -1$ when $p = 5, 7, 17, 19, 29, 31, 41, 43, 53, 67, 79$, and 89.

Now you *really* need to be told there is a pattern before you will spot it. The answer is:

$$\left(\frac{3}{p}\right) = \begin{cases} 1 & \text{if } p \equiv 1 \pmod{12} \\ -1 & \text{if } p \equiv 5 \pmod{12} \\ -1 & \text{if } p \equiv 7 \pmod{12} \\ 1 & \text{if } p \equiv 11 \pmod{12}. \end{cases}$$

We would like to go a bit further along. There is no point in computing $\left(\frac{4}{p}\right)$, because $x^2 - 4 \equiv 0$ always has two solutions.

EXERCISE: What are the two solutions of $x^2 - 4 \equiv 0 \pmod{p}$?

SOLUTION: One part is easy: $x \equiv 2 \pmod{p}$ is a solution. The other solution may be more difficult to spot. You need to recall our observation from earlier that the two elements of $S(\mathbf{F}_p)$ will add up to p. Sure enough, you can try plugging $p - 2$ into $x^2 - 4$, and $(p-2)^2 - 4 \equiv (-2)^2 - 4 \equiv 0 \pmod{p}$. Or, in terms of $\mathbf{F}_p$, we can just say that the two solutions are 2 and -2.

Remember that when we count solutions, we are counting them as elements of $\mathbf{F}_p$, so that all possibilities that are congruent modulo p count as just one solution. For instance, if $p = 7$, then $2^2 \equiv 9^2 \equiv 16^2 \equiv -5^2 \equiv -12^2 \ldots \pmod{7}$, but that is just a single solution to the equation $x^2 - 4 = 0$ in $\mathbf{F}_7$.

When Is 5 a Square mod p? (Will This Go On Forever?)

Now we try $a = 5$:

EXERCISE: Compute $\left(\frac{5}{p}\right)$ for all odd primes under 100.

SOLUTION: Now we work out that $\left(\frac{5}{p}\right) = 1$ when $p = 11$, 19, 29, 31, 41, 59, 61, 71, 79, and 89, and $\left(\frac{5}{p}\right) = -1$ when $p = 3$, 7, 13, 17, 23, 37, 43, 47, 53, 67, 73, 83, and 97.

Perhaps you are starting to spot the pattern. This would be much easier if we used all primes up to 1,000. The difficulty of spotting the pattern is eased slightly because $p = 5$, so you can look at the last digit of each prime on each list, and use that to guide your guess. One pattern is that

$$\left(\frac{5}{p}\right) = \begin{cases} 1 & \text{if } p \equiv 1 \pmod 5 \\ -1 & \text{if } p \equiv 2 \pmod 5 \\ -1 & \text{if } p \equiv 3 \pmod 5 \\ 1 & \text{if } p \equiv 4 \pmod 5. \end{cases}$$

That answer is misleading. The form of the answer that we will find most helpful[5] is that

$$\left(\frac{5}{p}\right) = \begin{cases} 1 & \text{if } p \equiv 1 \pmod{20} \\ -1 & \text{if } p \equiv 3 \pmod{20} \\ -1 & \text{if } p \equiv 7 \pmod{20} \\ 1 & \text{if } p \equiv 9 \pmod{20} \\ 1 & \text{if } p \equiv 11 \pmod{20} \\ -1 & \text{if } p \equiv 13 \pmod{20} \\ -1 & \text{if } p \equiv 17 \pmod{20} \\ 1 & \text{if } p \equiv 19 \pmod{20}. \end{cases}$$

[5]We will see later in this chapter that we want the modulus in these patterns always to be a multiple of 4.

The Law of Quadratic Reciprocity

You do not need to do the work, but here is the corresponding table for $a = 7$:[6]

$$\left(\frac{7}{p}\right) = \begin{cases} 1 & \text{if } p \equiv 1 \pmod{28} \\ 1 & \text{if } p \equiv 3 \pmod{28} \\ -1 & \text{if } p \equiv 5 \pmod{28} \\ 1 & \text{if } p \equiv 9 \pmod{28} \\ -1 & \text{if } p \equiv 11 \pmod{28} \\ -1 & \text{if } p \equiv 13 \pmod{28} \\ -1 & \text{if } p \equiv 15 \pmod{28} \\ -1 & \text{if } p \equiv 17 \pmod{28} \\ 1 & \text{if } p \equiv 19 \pmod{28} \\ -1 & \text{if } p \equiv 23 \pmod{28} \\ 1 & \text{if } p \equiv 25 \pmod{28} \\ 1 & \text{if } p \equiv 27 \pmod{28}. \end{cases}$$

It can be difficult to see the forest for the trees here. One pattern that remains the same from chart to chart is that when we are computing $\left(\frac{a}{p}\right)$, the modulus in the chart is $4a$. Moreover, the charts are all symmetric. This can be expressed as follows:

THEOREM 7.2: Let a be a positive integer.

1. If p and q are two odd primes so that $p \equiv q \pmod{4a}$, then $\left(\frac{a}{p}\right) = \left(\frac{a}{q}\right)$.
2. If p and q are two odd primes so that $p + q \equiv 0 \pmod{4a}$, then $\left(\frac{a}{p}\right) = \left(\frac{a}{q}\right)$.

Theorem 7.2 is a surprisingly difficult statement to prove (even if we just pick $a = 3$). Like most such statements, it has amazing

[6] Can you explain why there is no need to try $a = 6$?

consequences. With it, we can finally get around to explaining the title of this chapter, the Law of Quadratic Reciprocity:

THEOREM 7.3 (*Quadratic Reciprocity*): Suppose that p and q are odd primes.

1.

$$\left(\frac{-1}{p}\right) = \begin{cases} 1 & \text{if } p \equiv 1 \pmod{4} \\ -1 & \text{if } p \equiv 3 \pmod{4}. \end{cases} \tag{7.4}$$

2.

$$\left(\frac{2}{p}\right) = \begin{cases} 1 & \text{if } p \equiv 1 \text{ or } 7 \pmod{8} \\ -1 & \text{if } p \equiv 3 \text{ or } 5 \pmod{8}. \end{cases} \tag{7.5}$$

3. If $p \equiv q \equiv 3 \pmod{4}$, then

$$\left(\frac{p}{q}\right) = -\left(\frac{q}{p}\right). \tag{7.6}$$

4. If p or q or both are $\equiv 1 \pmod{4}$, then

$$\left(\frac{p}{q}\right) = \left(\frac{q}{p}\right). \tag{7.7}$$

The format of equations (7.6) and (7.7) is why the theorem is called a *reciprocity law*. The numbers p and q play reciprocal roles with respect to each other in these two equations. Although we will not prove Theorem 7.2 in this book, the proof of (7.6) and (7.7)—*assuming* the truth of Theorem 7.2—is just a complicated bit of algebraic manipulation. Feel free to skip it if you like. Here it is:

PROOF: First, we suppose that $p \equiv q \equiv 3 \pmod{4}$. Assume that $p > q$, and write $p = q + 4n$. Then $\left(\frac{p}{q}\right) = \left(\frac{q+4n}{q}\right) = \left(\frac{4n}{q}\right)$. Next, we use (7.1) from page 73 to rewrite $\left(\frac{4n}{q}\right)$ as $\left(\frac{4}{q}\right)\left(\frac{n}{q}\right)$, and because $\left(\frac{4}{q}\right) = 1$, we have now proved that $\left(\frac{p}{q}\right) = \left(\frac{n}{q}\right)$.

Now we use Theorem 7.2. Because $p = q + 4n$, we have $p \equiv q \pmod{4n}$, and so $\left(\frac{n}{q}\right) = \left(\frac{n}{p}\right)$. Because $p \equiv 3 \pmod 4$, we have $-\left(\frac{-1}{p}\right) = 1$. Multiplying by this disguised form of the number 1, we get $\left(\frac{n}{p}\right) = -\left(\frac{-1}{p}\right)\left(\frac{n}{p}\right) = -\left(\frac{-n}{p}\right)$. Next, we multiply by 1 again, this time disguised as $\left(\frac{4}{p}\right)$, to get $\left(\frac{n}{p}\right) = -\left(\frac{-n}{p}\right)\left(\frac{4}{p}\right) = -\left(\frac{-4n}{p}\right)$. And now, finally, we're done: $-\left(\frac{-4n}{p}\right) = -\left(\frac{p-4n}{p}\right) = -\left(\frac{q}{p}\right)$.

If $p \equiv q \equiv 1 \pmod 4$, the same proof works, except that the disguised form of the number 1 is now $\left(\frac{-1}{p}\right)$.

Finally, suppose that $p \equiv 1 \pmod 4$ and $q \equiv 3 \pmod 4$. Then $p + q$ is a multiple of 4, and now we write $p + q = 4n$. We start as before: $\left(\frac{p}{q}\right) = \left(\frac{4n-q}{q}\right) = \left(\frac{4n}{q}\right) = \left(\frac{4}{q}\right)\left(\frac{n}{q}\right) = \left(\frac{n}{q}\right)$. In exactly the same way, $\left(\frac{q}{p}\right) = \left(\frac{n}{p}\right)$. Now, we use the second piece of Theorem 7.2. Because $p + q \equiv 0 \pmod{4n}$, we have $\left(\frac{n}{q}\right) = \left(\frac{n}{p}\right)$. Combine the pieces, and we have $\left(\frac{p}{q}\right) = \left(\frac{q}{p}\right)$. □

Examples of Quadratic Reciprocity

Now that we have gone to such great lengths to prove Theorem 7.3, we should show you an example or two of its use.

EXERCISE: Compute $\left(\frac{24}{31}\right)$.

SOLUTION: We mentioned above a fact that is critical in using quadratic reciprocity to speed the computation of

Legendre symbols. Expressed symbolically, it is the statement that if $a \equiv b \pmod{p}$, then $\left(\frac{a}{p}\right) = \left(\frac{b}{p}\right)$. If you think back to the definition of the Legendre symbol, you will see why this is true. The symbol $\left(\frac{a}{p}\right)$ tells you something about solving the congruence $x^2 \equiv a \pmod{p}$. If $a \equiv b \pmod{p}$, then solving $x^2 \equiv a \pmod{p}$ is exactly the same as solving the congruence $x^2 \equiv b \pmod{p}$.

We can start our computation of $\left(\frac{24}{31}\right)$ by factoring $24 = 4 \cdot 6$. By property (7.1), we can write

$$\left(\frac{24}{31}\right) = \left(\frac{4}{31}\right)\left(\frac{6}{31}\right).$$

We know right away that $\left(\frac{4}{31}\right) = 1$, because 4 is a perfect square. So we can conclude that $\left(\frac{24}{31}\right) = \left(\frac{6}{31}\right)$.

Now, factor $6 = 2 \cdot 3$, and we get

$$\left(\frac{24}{31}\right) = \left(\frac{6}{31}\right) = \left(\frac{2}{31}\right)\left(\frac{3}{31}\right).$$

We use (7.5): Because $31 \equiv 7 \pmod{8}$, we know that $\left(\frac{2}{31}\right) = 1$. So now we know that $\left(\frac{24}{31}\right) = \left(\frac{3}{31}\right)$.

Next, we use (7.6), because $31 \equiv 3 \pmod{4}$ and $3 \equiv 3 \pmod{4}$, so

$$\left(\frac{3}{31}\right) = -\left(\frac{31}{3}\right).$$

Because $31 \equiv 1 \pmod{3}$, we know that $\left(\frac{31}{3}\right) = \left(\frac{1}{3}\right)$, which we know must be 1 (because 1 is a perfect square). Put all of this together, and we have worked out that $\left(\frac{24}{31}\right) = -1$, without solving any quadratic equations.

We now know that there is no integer a whose square leaves a remainder of 24 when divided by 31. Of course, it would not be hard simply to check all a's from 0 to 30 and see whether $a^2 \equiv 24 \pmod{31}$. But when the modulus is very big, it is much faster to use quadratic reciprocity. If the modulus is very very big, or is not specified (it might just be "any p" appearing in the proof of some theorem), then we might *have* to use quadratic reciprocity.

Here is a medium-sized numerical example:

EXERCISE: Using the Law of Quadratic Reciprocity, compute $\left(\frac{3{,}411}{3{,}457}\right)$.

SOLUTION: Again, we start with a factorization: $3{,}411 = 3^2 \cdot 379$. So we have

$$\left(\frac{3{,}411}{3{,}457}\right) = \left(\frac{9}{3{,}457}\right)\left(\frac{379}{3{,}457}\right) = \left(\frac{379}{3{,}457}\right).$$

Now, $379 \equiv 3 \pmod 4$ and $3457 \equiv 1 \pmod 4$, so we use (7.7) and get

$$\left(\frac{379}{3{,}457}\right) = \left(\frac{3{,}457}{379}\right).$$

We continue by noticing that $3{,}457 \equiv 46 \pmod{379}$, and so

$$\left(\frac{3{,}457}{379}\right) = \left(\frac{46}{379}\right) = \left(\frac{2}{379}\right)\left(\frac{23}{379}\right).$$

We use (7.5) to compute $\left(\frac{2}{379}\right) = -1$, because $379 \equiv 3 \pmod 8$. So now we have the equation $\left(\frac{3{,}411}{3{,}457}\right) = -\left(\frac{23}{379}\right)$.

Next, we see that $23 \equiv 3 \pmod 4$ and $379 \equiv 3 \pmod 4$, and so by (7.6) we get

$$-\left(\frac{23}{379}\right) = \left(\frac{379}{23}\right).$$

Because $379 \equiv 11 \pmod{23}$ we get

$$\left(\frac{379}{23}\right) = \left(\frac{11}{23}\right)$$

and then we use (7.6) again to see that

$$\left(\frac{11}{23}\right) = -\left(\frac{23}{11}\right) = -\left(\frac{1}{11}\right) = -1.$$

So in the end, we have computed $\left(\frac{3{,}411}{3{,}457}\right) = -1$, again without having to solve any quadratic equations.

We will return to this subject later, after we have explained more about representations, and we will explain why the Law of Quadratic Reciprocity can be viewed as an example in representation theory. We will also present some other examples of reciprocity laws and explain why they are indeed reciprocity laws. Finally, we will give some instances of how reciprocity laws can be used to study solution sets of more complicated Diophantine equations.

PART TWO

Galois Theory and Representations

Chapter 8

GALOIS THEORY

Road Map

We now define and explain the central actor of our drama. The "absolute Galois group of **Q**" is the set of all permutations of roots of **Z**-polynomials that preserve addition and multiplication. We will call it "the Galois group" for short and denote it by "G," because we believe it is the group par excellence.

Knowledge about G helps us to solve systems of **Z**-equations. Curiously, the deeper our knowledge grows, the more our interest shifts to G itself, away from particular equations. The big news is when G is used to solve an old problem, such as Fermat's Last Theorem. But the "story behind the news" is G itself, whose shadowy powers are still being slowly discovered by number theorists.

G is a group of symmetries and has a rich structure—much of it still unknown. One of the best ways of probing that structure is to study the representations of G into standard objects, namely, permutations groups and matrix groups. Following this chapter and the next, which provide two deep examples of G's activity—namely, the roots of **Z**-polynomials and the torsion points of elliptic curves—we will introduce matrices, matrix groups, and group representations. Then we will have most of the ingredients we need to finish our story.

Polynomials and Their Roots

We review here the basic facts and definitions about polynomials.

1. A **Z**-polynomial is a polynomial in one variable with integer coefficients.
2. The *degree* of a **Z**-polynomial is the degree of the highest power of the variable that occurs. For example, the degree of $11x^5 + 103x - 41$ is 5.
3. Any **Z**-polynomial f with degree at least 1 can be factored into a nonzero constant c times the product of factors of the form $(x - a)$, where a is some complex number.
4. If the degree of f is d, then there are exactly d factors.
5. The *roots* of f are defined to be those complex numbers b such that $f(b) = 0$. We can think of the equation $f(x) = 0$ as defining a variety, S_f, and then the roots of the polynomial are the elements of $S_f(\mathbf{C})$.

The roots of f are exactly the a's appearing in its factorization into terms of the form $(x - a)$. This is easy to prove. Suppose that we have factored

$$f(x) = c(x - a_1)(x - a_2)\cdots(x - a_d).$$

The number of factors on the right-hand side, d, must be the same as the degree of $f(x)$. If we replace x by any of the numbers $a_1, a_2, \ldots, a_d$, throughout the equation, the right-hand side multiplies out to be 0, so the left-hand side must also be 0; that is, $f(a_1) = f(a_2) = \cdots = f(a_d) = 0$. On the other hand, if we substitute in some number t which is different from all the a_k's, then the right-hand side is a product of nonzero numbers, and so must be nonzero. This says that $f(t) \neq 0$. Thus, we have shown that $f(t) = 0$ if and only if t is one of the numbers a_k.

Before we give some examples, notice that a factor of the form $(x + b)$ is really of the form $(x - a)$ where $a = -b$. For instance, $(x + 3) = (x - (-3))$. So when we factor $f(x)$ into linear factors, we will feel free to use factors of either form: $(x - a)$ or $(x + b)$.

EXAMPLE: Let $f(x) = x^2 - x - 1$. Then $f(x)$ factors as $f(x) = (x - \phi)(x - \phi')$ where $\phi = (1 + \sqrt{5})/2 = 1.6180340\ldots$ and $\phi' = (1 - \sqrt{5})/2 = -0.6180340\ldots$. So the roots of $f(x)$ are ϕ and ϕ'. By the way, ϕ is an interesting number called the "golden ratio." It is supposed to describe a particularly pleasing proportion that occurs in nature. A recent book on the subject is (Livio, 2002).

EXAMPLE: Let $f(x) = 5(x - 1)(x + 2)(x - 3)$. Here we began with $f(x)$ already factored, so we can see that its roots are 1, -2, and 3. We can multiply out f and see it explicitly as a third-degree polynomial: $f(x) = 5x^3 - 10x^2 - 25x + 30$.

EXAMPLE: Consider $f(x) = x^4 - 1$. We expect four roots, but to see them all we have to use the complex numbers. The roots are 1, -1, i, and $-i$, and this corresponds to the factorization $f(x) = (x - 1)(x + 1)(x - i)(x + i)$.

The Field of Algebraic Numbers $\mathbf{Q}^{\text{alg}}$

DEFINITION: A complex number is said to be *algebraic* if it is the root of some **Z**-polynomial. The set of all algebraic numbers is denoted by $\mathbf{Q}^{\text{alg}}$.

The set $\mathbf{Q}^{\text{alg}}$ contains every integer n, because n is a root of the equation $x - n = 0$. It also contains every fraction a/b, because a/b is the solution of $bx - a = 0$. The set contains square roots of every fraction a/b, because $\sqrt{a/b}$ solves the equation $bx^2 - a = 0$. (Notice in particular that this means $\mathbf{Q}^{\text{alg}}$ contains i, the square root of -1, because it is a solution of $x^2 + 1 = 0$.) In fact, the set $\mathbf{Q}^{\text{alg}}$ contains nth roots of every fraction, because $\sqrt[n]{a/b}$ solves the equation $bx^n - a = 0$.

But there are many more complicated types of algebraic numbers. This is partly because the sum, difference, product, and quotient of any two algebraic numbers will also be algebraic. Therefore, $\mathbf{Q}^{\text{alg}}$ is a field. This assertion is a difficult theorem to prove, which we will not do in this book. As an example, you can try the following exercise:

> **EXERCISE**: Find a **Z**-polynomial that has a root $\sqrt{2} + \sqrt[3]{5}$.
>
> **SOLUTION**: $x^6 - 6x^4 - 10x^3 + 12x^2 - 60x + 17$. To find this solution, let $x = \sqrt{2} + \sqrt[3]{5}$. Then $(x - \sqrt{2})^3 = 5$. Multiply out the left-hand side, and you get $x^3 - 3\sqrt{2}x^2 + 6x - 2\sqrt{2} = 5$. Now rearrange this to be $x^3 + 6x - 5 = 3\sqrt{2}x^2 + 2\sqrt{2}$, square both sides, and collect terms.

The exercise illustrates the following fact, which we will not prove:

> **THEOREM 8.1**: If α is an algebraic number that is a root of a **Z**-polynomial of degree m, and β is an algebraic number that is a root of a **Z**-polynomial of degree n, then $\alpha + \beta$, $\alpha - \beta$, $\alpha\beta$ and α/β (if $\beta \neq 0$) are algebraic numbers, each of which is a root of some **Z**-polynomial of degree no more than mn.

More examples: The number π is *not* in $\mathbf{Q}^{\text{alg}}$, and this fact is very difficult to prove. The nineteenth-century mathematician Georg Cantor proved that "almost all" real numbers are not in $\mathbf{Q}^{\text{alg}}$, but it can be difficult to decide whether a particular number is or is not in $\mathbf{Q}^{\text{alg}}$.

One interesting question: Can every algebraic number be obtained, starting from the integers, by the repeated operations of addition, subtraction, multiplication, division, and $\sqrt[n]{\ }$ for various n? The well-known answer is: No. For example, let w be a root of the following **Z**-polynomial of degree 5: $2x^5 - 10x + 5$. The Norwegian mathematician Niels Abel proved that w could not be expressed starting only with integers and using just addition, subtraction, multiplication, division, and $\sqrt[n]{\ }$.

There are lots of algebraic numbers, and they are connected in many complicated ways. To start with, we have already stated the fact that if you add, subtract, multiply or divide any two algebraic numbers (except, of course, you cannot divide by 0), the answer is again an algebraic number. It is also true that if you take a square root, or cube root, or any other nth root of an algebraic number, you get an algebraic number. In fact, if you take a polynomial whose coefficients are *arbitrary algebraic numbers*, then it too can be factored completely into a constant c times the product of factors of the form $(x - a)$, where c and the a's are still algebraic numbers. In other words:

THEOREM 8.2: Suppose that $f(x) = a_n x^n + \cdots + a_1 x + a_0$, where the numbers $a_0, a_1 \cdots, a_n$ are elements of $\mathbf{Q}^{\text{alg}}$. Then all of the roots of $f(x)$ are also elements of $\mathbf{Q}^{\text{alg}}$.

The experts sum up Theorems 8.1 and 8.2 by saying that $\mathbf{Q}^{\text{alg}}$ is an *algebraically closed field*. It means that no matter how we play around with numbers, starting from the integers, forming polynomial equations, solving them, taking the solutions, forming more polynomial equations using those roots as coefficients, solving them, "etc. etc. etc."[1] we keep coming up with only algebraic numbers.

Of course, if we try to do something nonalgebraic, such as take the sine or cosine of an algebraic number, or raise one algebraic number to the power of another algebraic number, then we may get something nonalgebraic. So we forbid these operations.

You might ask: "Why don't we call a^b an algebraic operation, where a and b can be any numbers at all, not just fractions?" For example, $\sqrt{2}^{\sqrt{3}}$ is not considered algebraic. The reason is that we reserve the term *algebraic* for what can be derived by solving equations expressed in terms solely of the four operations of addition, subtraction, multiplication, and division. Raising to an integer power is acceptable because that is multiplying or dividing a repeated number of times. Even raising to a fractional power is acceptable, because taking nth roots is finding a number whose nth power is equal to the base. But to define $\sqrt{2}^{\sqrt{3}}$, we have to write $\sqrt{3}$ as an infinite, nonrepeating decimal, truncate it at the thousandth decimal place, raise $\sqrt{2}$ to that truncated fractional power, then do this again for the millionth decimal place, "etc. etc. etc.," and in

[1]To quote Gibbon, *Decline and Fall of the Roman Empire*, Chapter XX, footnote 29.

the limit, you get the value of $\sqrt{2}^{\sqrt{3}}$. Because we have to perform a limiting process, this innocuous expression $\sqrt{2}^{\sqrt{3}}$ really belongs to calculus and not to algebra.[2]

The Absolute Galois Group of Q Defined

How do we bring order to what seems to be a chaos of algebraic numbers? One could paraphrase Alexander Pope's couplet on Isaac Newton and say: God said, "Let Galois be." But that could be overstating the case. There is still a lot of chaos left over, even after we understand Galois theory. However, Galois theory is what gives us a way to ask intelligent questions about this swirling mass of algebraic numbers—that, together with representation theory.

We begin by defining the "absolute Galois group of $\mathbf{Q}$." It is usually denoted by the symbol $G_{\mathbf{Q}}$. But for short, we will call it "*the* Galois group" and usually denote it simply by the letter G. It is called "absolute" to distinguish it from the various "relative" Galois groups of polynomials that will be introduced in chapter 13.[3]

> **DEFINITION**: The *absolute Galois group* G is made up of all permutations g of $\mathbf{Q}^{\text{alg}}$ that preserve addition and multiplication. That is, for any numbers a and b in $\mathbf{Q}^{\text{alg}}$, we must have $g(a+b) = g(a) + g(b)$ and $g(ab) = g(b)g(b)$.

Here is a brief dialogue that will help to explain this definition. First, notice that because $\mathbf{Q}^{\text{alg}}$ is an infinite set, its group of permutations is also infinite. But we do not let just *any* permutation of $\mathbf{Q}^{\text{alg}}$ into G.

---

[2]The reader might not be satisfied with this discussion. After all, $\sqrt{2}$ can be defined by a limiting process as well, and yet we are allowing $\sqrt{2}$ as an element of $\mathbf{Q}^{\text{alg}}$. The point is that there *are* ways to define $\sqrt{2}$ without resort to a limiting process, namely, as the positive root of $x^2 - 2$, while a difficult theorem asserts that there is no way to define $\sqrt{2}^{\sqrt{3}}$ without using a limiting process.

[3]This would be a good time to review the concept of "permutation" from chapter 3, if you need to. Remember that a permutation of a set A is a function from A to itself, which is a one-to-one correspondence.

A Conversation with s: A Playlet in Three Short Scenes

Meet the permutation s:

—Hello. Pleased to meet you. So you are a one-to-one correspondence between $\mathbf{Q}^{\mathrm{alg}}$ and itself? And you would like to join the G-club? I'm afraid there are some tests you will have to pass before we can admit you.

—Please, I'm ready to try.

—First, let's check that you really are a permutation of $\mathbf{Q}^{\mathrm{alg}}$. If I give you an element a of $\mathbf{Q}^{\mathrm{alg}}$, that is to say, if I give you an algebraic number a, what will you do with it?

—I'll take it and output another algebraic number, let's say b. Only one output. I'm never in doubt. And I always turn a into b. I'm an honest function.

—Are you sure you won't ever output π or something nonalgebraic like that?

—I won't.

—Good. And would you ever give the *same* output for two different algebraic number inputs?

—No.

—That's good too. And might there be some algebraic number that is never an output of yours?

—No. If you name any algebraic number at all, for example $\sqrt{2} + \sqrt[5]{11}$, then I can find some algebraic number input whose output it would be.

—Very well, so you are a permutation. But that's still very far from what you need to enter G. So far, all I've done is checked your status as a permutation. The big test lies ahead. Let's go slowly. What do you do with 0 as input?

—I output 0.

—Good. What do you do with 1?

—I output 1.

(This could go on quite a while, so we will summarize a bit. We'll say that s "sends" a to b if for the input a she outputs b. In functional notation we could write $b = s(a)$. It turns out that s sends every rational number, that is, every fraction $\frac{c}{d}$, where c and d are integers, to itself. That is good, because we will prove later in this chapter that s must have this property in order to belong to the Galois group G.)

—Maybe you are just the identity permutation. Do you in fact send every algebraic number to itself?

—No, I'm more interesting than that.

—Good, glad to hear that. The identity permutation is already a member of the G-club anyway. Now let's get down to business. What do you do with $\sqrt{2}$?

—I send it to $-\sqrt{2}$.

❄❄

(*Later that afternoon.*)

—What do you do with a root of $1 + x^2 + 33x^3 - x^7 + 4x^9$?

—I send it to another root of the same polynomial $1 + x^2 + 33x^3 - x^7 + 4x^9$.

—Hmm. Is that a general rule? If I give you a root of any **Z**-polynomial, will you always send it to another root, either the same one or a different one, of the *same* **Z**-polynomial?

—(*proudly*) Yes, I will.

—Good, things look hopeful. But now we have to come down to brass tacks. Do you preserve all arithmetic operations: addition, subtraction, multiplication, and division?

—Excuse me, I'm not sure what you mean.

—Well, for example, suppose you send a to c and b to d. Tell me what you will do with $a + b$.

—I'll send it to $c + d$.

—Are you sure? Always?

—Yes. For example, because I send $\sqrt{3}$ to itself—$\sqrt{3}$—and I send $\sqrt{2}$ to $-\sqrt{2}$, then I will send $\sqrt{3} + \sqrt{2}$ to . . . let me check . . . yes, to $\sqrt{3} - \sqrt{2}$, as I claimed. Also, because I send 5 to itself and $\sqrt{2}$ to $-\sqrt{2}$, then I will send $5 + \sqrt{2}$ to $5 - \sqrt{2}$. And so on.

—What about $a - b$, ab and $\frac{a}{b}$?

—I'll send them to $c - d$, cd and $\frac{c}{d}$, without fail.

—Very good, you pass all the tests and can join the club. You are now a member of G.

(Later, in the club dining room.)

—Just out of curiosity, what is your name?

—My name is s.

—An interesting name. And tell me, what would you do with $\sqrt[4]{2}$?

—Well, $\sqrt[4]{2}$ is the square root of the square root of 2, namely $\sqrt{\sqrt{2}}$. By the way, whenever I use the symbol $\sqrt{\ }$ with a positive number, I intend the positive square root. So, because I send $\sqrt{2}$ to $-\sqrt{2}$, I will have to send $\sqrt{\sqrt{2}}$ to $\sqrt{-\sqrt{2}}$, and that's the square root of a negative number. The answer will be an imaginary number. Let's refer to $\sqrt[4]{2}$ by the Greek letter "Ψ" for short. Then, I have to send Ψ to either $i\Psi$ or $-i\Psi$. I'll let you guess which one....

(Here s smiles enigmatically and the record of the conversation breaks off.)

We can summarize as follows: The group G consists of permutations of $\mathbf{Q}^{\text{alg}}$. Not every permutation is allowed, but only and precisely those permutations that preserve the four basic arithmetic operations: addition, subtraction, multiplication, and division. We can write this in terms of equations like this: If g is an element of G, and if a and b are any numbers in $\mathbf{Q}^{\text{alg}}$, then $g(a)$ is the number that g sends a to, and $g(b)$ is the number that g sends b to. Then

$$g(a+b) = g(a) + g(b). \tag{8.3}$$

$$g(a-b) = g(a) - g(b). \tag{8.4}$$

$$g(ab) = g(a)g(b). \tag{8.5}$$

$$g\left(\frac{a}{b}\right) = \frac{g(a)}{g(b)}, \text{ if } b \neq 0.^4 \tag{8.6}$$

[4] In fact, conditions (8.4) and (8.6) are redundant in this definition. They are implied by conditions (8.3) and (8.5) and the requirement that g is a permutation. It is not too hard to prove this assertion.

Digression: Symmetry

A symmetry is a function that preserves what we feel is important about an object. For example, a starfish has fivefold symmetry, which means there is a function (rotation by 72°) that we can perform on the starfish which keeps it looking pretty much the same. (*Fivefold* means that if we perform this function five times we will come back to the original state.) Humans have twofold symmetry, because if we reflect them in a mirror, they still look pretty much the same. If we rotate humans 180°, they will be standing on their heads or facing away from the mirror (or both). So reflection in a mirror is a different kind of symmetry from rotation. But it is twofold because if we reflect the reflection we get back the original.

These symmetries are not exact, for nothing in living nature is absolutely exact. But in mathematics we can study exact symmetries.[5] In our case, each g in the Galois group G is a symmetry of $\mathbf{Q}^{\text{alg}}$ because it preserves the operations that concern us in algebra, namely, addition and multiplication. This use of the word "symmetry" is part of the justification for the title of this book.

How Elements of G Behave

The four equations (8.3)–(8.6) might seem like no big deal, but they have many consequences and put many constraints on the permutation g if it is to be allowed into G. Some of these constraints are obvious and some of them are very subtle—so subtle that Galois theory still possesses a large realm of mystery.

Let us look at some of the less subtle constraints on g. First, if we set $a = b$, then from equation (8.4) we derive $g(0) = 0$, and if we set $a = b = 1$, we can then derive from equation (8.5) that $g(1) = 1$.[6] So g must send 0 to 0 and 1 to 1. This explains the interviewer's first two questions.

---

[5] A classic reference about symmetries is (Weyl, 1989).

[6] Because g is a permutation and $g(0) = 0$, it follows that $g(1) \neq 0$, so it can be cancelled on both sides of the equation $g(1) = g(1)g(1)$.

Now let $a = 1$ and $b = 1$ and use equation (8.3). We obtain $g(2) = 2$. So g must send 2 to 2. Do you see the pattern? Setting $a = 1$ and $b = 2$, we can now use equation (8.3) again to see that g must send 3 to 3. And so on. We conclude that if a is any positive integer, then $g(a) = a$.

Notice that this argument is a kind of bootstrap. We cannot show directly that $g(13) = 13$, but we have to build it up slowly, first 1, then 2, then 3, and so on, until we get to 13. Slow but sure wins the race—we can eventually bootstrap ourselves up to any positive integer we want, and so we have justified the statement that $g(a) = a$ for every positive integer a. This kind of argument has a formal name: *mathematical induction*.

EXERCISE: Use equation (8.4) over and over again, starting with $a = 0$ and $b = 1$, to show that $g(k) = k$ for every negative integer k.

So we have seen that if g is in G, then it takes every integer to itself. But now if we have the fraction $\frac{a}{b}$, where b is not zero, and a and b are integers, we can use formula (8.6) and derive that $g(\frac{a}{b}) = \frac{g(a)}{g(b)} = \frac{a}{b}$. In other words, g takes every rational number to itself. This is why the interviewer kept asking this type of question.

Perhaps you are beginning to suspect that g must take *every* algebraic number to itself, so that G would contain only one element, the identity permutation, which we shall call "e." In this case, G would not be very interesting and we would not be talking about it. But because you have met s, you know that there are other elements in G. Our acquaintance s, for example, takes $\sqrt{2}$ to $-\sqrt{2}$. In fact, G has infinitely many elements.

You may want us to describe in some thorough way an element in G that is not the identity permutation. But here is an amazing fact, one of the facts that make Galois theory so hard: There is only one element of G, other than the neutral element e, for which we can give a complete description that would satisfy you. This element is called "c" and we will describe it in the next paragraph. But except for e and c, *no other element* of G can be written down explicitly.

Instead, we have an "existence theorem" that shows us how to be sure there are lots of elements in G, and allows us to know a certain amount of information about each one explicitly, but not a *complete* description of any of them, except e and c.

What is c? Its full name is *complex conjugation*. We used it before on page 46, when we were taking the reciprocal of a complex number. Remember that $\mathbf{Q}^{\text{alg}}$ contains various complex numbers. Every complex number is of the form $x+yi$, where x and y are real numbers.[7] By definition $c(x+yi)=x-yi$. You can carefully check that c satisfies all the requirements (8.3)–(8.6) to be in G.

Before we try to explain the existence theorem mentioned above, we have to learn more about G and about $\mathbf{Q}^{\text{alg}}$. Here comes the key idea: Take a **Z**-polynomial $f(x)$. Remember that, by definition, every algebraic number is the root of some **Z**-polynomial (in fact, of many different ones).

THEOREM 8.7: Suppose g is an element of G, and suppose a is a root of the **Z**-polynomial $f(x)$. Then $g(a)$ is also a root of the same **Z**-polynomial, $f(x)$. In other words, if $f(a)=0$ then $f(g(a))=0$.

This is extremely important. To understand what it is saying, we look at an example. Suppose $f(x)=x^2-2$. Then $f(x)$ has exactly two roots, $\sqrt{2}$ and $-\sqrt{2}$. Therefore, if g is any element at all of G, then g must send $\sqrt{2}$ to one of these two roots, either to itself—$\sqrt{2}$—or to the other root: $-\sqrt{2}$. Do you remember that s sent $\sqrt{2}$ to $-\sqrt{2}$? There wasn't much choice. She had only two choices. Half of the elements of G send $\sqrt{2}$ to $\sqrt{2}$ and the other half send it to $-\sqrt{2}$.

Now consider what further choices an element g of G must make. She has just chosen whether to send $\sqrt{2}$ to $\sqrt{2}$ or to $-\sqrt{2}$. And along comes the polynomial x^2-3, with *its* two roots $\sqrt{3}$ and $-\sqrt{3}$. So g must choose whether to send $\sqrt{3}$ to itself or to minus itself. You can see that there will be an infinite number of choices yet to be made. Fortunately this is mathematics, so there is no time limit—all the

[7] In fact, if $x+yi$ is a complex number in $\mathbf{Q}^{\text{alg}}$, then x and y each separately must be in $\mathbf{Q}^{\text{alg}}$.

choices have been made already in some Platonic heaven, and that gives g its unique selfhood as a permutation in G.

Notice that if (for example) g sends $\sqrt{2}$ to $-\sqrt{2}$, then because g is a permutation and must also send $-\sqrt{2}$ to a root of $x^2 - 2$, we can conclude that g *must* send $-\sqrt{2}$ to $\sqrt{2}$. Similarly, using formula (8.3) above, s knew where she had to send $\sqrt{3}+\sqrt{2}$ in her conversation with the club official.

So g does not always have to make a fresh decision where to send every algebraic number that comes knocking at her door. But in fact the subtlety consists precisely in this: After a few hundred decisions have been made, and some complicated number like $\sqrt{\sqrt{2}-\sqrt{3}+i\sqrt{6}}$ comes knocking, is its destination forced by the previous decisions or do we have some fresh decision to make?

Remember that any g in G must send any root of a **Z**-polynomial $f(x)$ to itself or to some other root of the *same* **Z**-polynomial. Why is this true? It actually follows easily from formulas (8.3)–(8.5). Evaluation of $f(a)$ means taking a and doing various multiplications of itself by itself, or by integers, and then adding or subtracting the answers. Because g sends each integer to itself and preserves addition, subtraction, and multiplication, we see that $f(g(a)) = g(f(a))$. This is true regardless of what value $f(a)$ happens to be. But if $f(a) = 0$, then we see that $f(g(a)) = g(0) = 0$.

As one last example for now, look at what complex conjugation c does to i: It sends i to $-i$. Is that a coincidence? No. The numbers i and $-i$ are the roots of the **Z**-polynomial $x^2 + 1$. So $c(i)$ had to be i or $-i$. As a matter of fact, $c(i) = -i$. (If $c(i)$ wanted to equal i, then c would not be c; it would be something else. But there are plenty of elements in G that send i to itself—in fact, half of them do.)

Now, picking a particular **Z**-polynomial gives us a useful way to think about an element g in G. Take any polynomial that you like with unequal roots, and list the roots: $a_1, a_2, \ldots, a_n$. We know that $g(a_1)$ must be some root of the same polynomial. Suppose that $g(a_1) = a_3$. Now let us move on to $g(a_2)$. It also has to be one of the roots. Because g is a one-to-one map, we know that $g(a_2)$ cannot be the same as $g(a_1)$. So we find out which of the remaining roots we get for $g(a_2)$, and then we move onto the next root.

When we are finished, we will discover that g is a permutation of the numbers $a_1, a_2, \ldots, a_n$. So all of the theorems that mathematicians have proved about permutations give us some insight into g. We can do this for lots of polynomials, and get lots of information.

One of the great mysteries of Galois theory is that there are certain polynomials with the property that, given a list of their roots $a_1, a_2, \ldots, a_n$, we can find an element g that permutes these roots *any way we like*. But for some polynomials, after we have chosen $g(a_1)$, we have no choice at all about the values of $g(a_2)$ or $g(a_3)$. Some of the time, we can choose $g(a_1)$ to be anything we want, and then $g(a_2)$ has a limited set of choices. After that, $g(a_3)$ will have no choices at all. There are examples of this kind of constraint in the Galois permutation of roots in chapters 13 and 18. As a general rule, it is not easy to look at a polynomial and know what permutations of its roots are or are not able to be produced by an element of G.

And there is something else interesting going on here. Rather than start with an element g in the mysterious group G, which has to decide what to do with *every* element of $\mathbf{Q}^{\text{alg}}$, we can instead pick one particular **Z**-polynomial, $f_1(x)$, and pick some permutation of the roots that is permissible. Then we can pick some other **Z**-polynomial, $f_2(x)$ and choose some permissible permutation of *those* roots, which is compatible[8] with the first set of choices that we made. We can then go on and on, picking polynomial after polynomial, and choosing permissible permutations of the roots, compatible with all previous choices. When we are all done—well, we will never be all done, because there are infinitely many **Z**-polynomials, but we can imagine being all done[9]—we will have found an element of G. The fact that this works and an infinite number of choices can all be made compatibly is the "existence

[8]Here, "compatible" means that as we choose the new permutation of the new roots and put it together with the old choices, requirements (8.3)–(8.6) on page 95 continue to be satisfied.

[9]Mathematicians use something called Zorn's Lemma here, to let us compress the process of doing infinitely many things into a short amount of time, so that we can still have time left over.

theorem” that we alluded to above. See chapter 14 for more on this theorem.

Why Is G a Group?

Finally, we need to check that the permutations in G form a group. That is, if g and h are in G, so that they permute the algebraic numbers $\mathbf{Q}^{\mathrm{alg}}$, then we must check that their composition as permutations, $g \circ h$, also permutes $\mathbf{Q}^{\mathrm{alg}}$ *and* satisfies the “rules of the club” (8.3)–(8.6) on page 95. This is not hard to do. We also must check that if g is in G, then its inverse permutation g^{-1} also abides by the rules of the club—which is also not hard to do. Easiest of all, we must check that the identity permutation e abides by these four rules. In other words, if g and h are in G, so are $g \circ h$, g^{-1}, and e. So G is a group. (The associative property of the group law still holds, because composition of permutations is always associative.)

Summary

We have covered an immense amount of ground in this chapter in order to describe the Galois group, or symmetry group, of the algebraic numbers. This is the key object we will be using as the source of the representations we want to discuss later. Since it is so key, here is a summary:

1. $\mathbf{Q}^{\mathrm{alg}}$ is the set of all complex numbers that can appear as roots of **Z**-polynomials.
2. The Galois group G is the set of all permutations of $\mathbf{Q}^{\mathrm{alg}}$ that preserve the operations of addition and multiplication.
3. If g is any element of G and $f(x)$ is any **Z**-polynomial, then as g acts as a permutation of $\mathbf{Q}^{\mathrm{alg}}$, it permutes the roots of $f(x)$. It never maps some root of $f(x)$ to a nonroot of $f(x)$.
4. G has infinitely many elements.

5. The only two elements of G that we can describe explicitly *in toto* are e, the identity permutation and c, complex conjugation.
6. Any element g of G can be partially described by taking a **Z**-polynomial $f(x)$, listing its roots $a_1, \dots, a_n$, and telling what permutation of these n algebraic numbers occurs when we apply g.
7. Zorn's Lemma plus some advanced algebra can be used to piece together the partial descriptions of item 6 to get elements of G. That is how we know G is infinite.

Chapter 9

ELLIPTIC CURVES

Road Map

We pause here to describe a class of **Z**-varieties called "elliptic curves." They are groups *and* varieties at the same time, which enables us to know much more about them than we do about most varieties. In this chapter, we only give the basic facts. In chapter 18, we will use them to obtain some beautiful and complicated representations of the absolute Galois group which were essential to the proof of Fermat's Last Theorem.

Elliptic Curves Are "Group Varieties"

Serge Lang began his book *Elliptic Curves: Diophantine Analysis* by writing: "It is possible to write endlessly about elliptic curves. (This is not a threat.)" That is because an elliptic curve is simultaneously an example of two different concepts that we have discussed in earlier chapters: varieties and groups. An elliptic curve E is the set of solutions to a certain kind of **Z**-equation so that for any field R, that is, any number system in which we can divide by any nonzero element, $E(R)$ is a group.[1]

Moreover, $E(R)$ is always an *abelian group*, which is the term that mathematicians give to a group in which the group law is commutative as well as associative. That is to say, in an abelian

---

[1]More exactly, $E(R)$ is a group after you extend it by adding one more element, called $\mathcal{O}$. And there is a slight condition on R that needs to be observed, as we shall see.

group $x \circ y = y \circ x$ for all elements x and y in the group. Because an elliptic curve is the set of solutions to a **Z**-equation, we can study how the Galois group permutes these solutions. And because it is an abelian group, we can apply the theorems of algebra about abelian groups. By the way, the name "elliptic curve" is derived from the fact that certain elliptic curves are connected to the problem of studying the arc-length of certain ellipses. An elliptic curve itself is *not* elliptical in shape.

To begin, an elliptic curve can be thought of as the set of solutions to the equation

$$y^2 = x^3 + Ax + B,$$

where A and B can be any two fixed integers, as long as $2(4A^3 + 27B^2) \neq 0$.[2] Usually, we use the letter E to stand for an elliptic curve.

We often use geometric terminology for the elements of $E(R)$, calling them "points." And we call the whole set $E(R)$ a "curve." This is because when R is the set of real numbers $\mathbf{R}$, we can actually graph $E(\mathbf{R})$ and view the solutions (x, y) as actual points on the resulting curve.

An Example

For example, E can be the elliptic curve defined by the equation $y^2 = x^3 + 1$ (so here we are choosing $A = 0$ and $B = 1$). The graph of $y^2 = x^3 + 1$ in the xy-plane is shown in Figure 9.1.

We view E as a variety, so, for example,

$$E(\mathbf{Z}) = \{(x, y) : y^2 = x^3 + 1, \quad x, y \in \mathbf{Z}\}$$

$$E(\mathbf{Q}) = \{(x, y) : y^2 = x^3 + 1, \quad x, y \in \mathbf{Q}\}$$

$$E(\mathbf{R}) = \{(x, y) : y^2 = x^3 + 1, \quad x, y \in \mathbf{R}\}$$

$$E(\mathbf{F}_5) = \{(x, y) : y^2 = x^3 + 1, \quad x, y \in \mathbf{F}_5\}$$

---

[2]The reason for this rather arcane restriction is that shortly, in one of our formulas, we might implicitly be dividing by a factor of $2(4A^3 + 27B^2)$, and we need to make sure that we do not divide by 0.

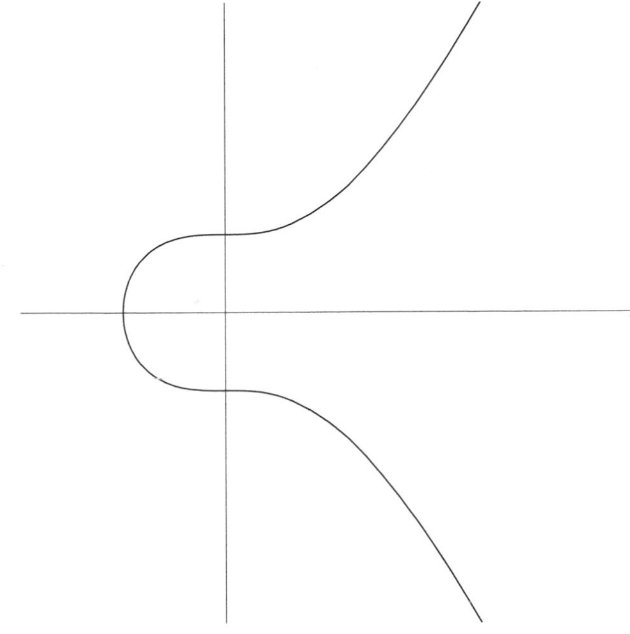

Figure 9.1: $y^2 = x^3 + 1$

and so on. In other words, $E(\mathbf{Z})$ tells us to search for *integers* that satisfy the defining equation for E, while $E(\mathbf{C})$ tells us to search for *complex numbers* that satisfy the same equation. However, when we plug a number system R into E to get $E(R)$, we are tacitly assuming that $2(4A^3 + 27B^2) \neq 0$ in R. So in our example we must assume that $6 \neq 0$ in R, so for instance $R = \mathbf{F}_2$ or $R = \mathbf{F}_3$ are not allowed for this E.

It is not so clear in this example what the elements of $E(\mathbf{Z})$ are. A bit of trial and error shows that if $x = 0$, then $y = \pm 1$ works, and if $x = -1$, then $y = 0$ also works. A bit more trial-and-error will let you work out that if $x = 2$, then $y = \pm 3$ works. It is even less obvious what the elements of $E(\mathbf{Q})$ are, or even if there are any elements of $E(\mathbf{Q})$ other than those in $E(\mathbf{Z})$.

On the other hand, it is a little easier to find out about $E(\mathbf{R})$: As long as we plug in a value of x that makes $x^3 + 1$ bigger than or equal to 0 (which means that we have to pick $x \geq -1$), then we can take a square root and find a y-value. For example, if we substitute in $x = 4$, we get $4^3 + 1 = 65$, and then $y = \sqrt{65}$ or $y = -\sqrt{65}$. This gives two different points on the elliptic curve: $(4, \sqrt{65})$ and $(4, -\sqrt{65})$.

It is even easier to search for elements of $E(\mathbf{C})$, because we can plug in *any* x-value at all and find the y-values. If $x^3 + 1$ is 0, i.e., if x is a cube root of -1, then there is only one suitable y-value, and otherwise there are two.

What about $E(\mathbf{F}_5)$? To begin with, because we are working with congruences, we must establish not just that $2(4A^3 + 27B^2) \neq 0$, but rather that $2(4A^3 + 27B^2) \not\equiv 0 \pmod 5$. We check that $2(4A^3 + 27B^2) = 54 \not\equiv 0 \pmod 5$, so we can continue. Now, the easiest thing to do is to try substituting in all five of the values $x = 0, 1, 2, 3$, and 4, and seeing if we can get a y-value or two that solve the equation. When we do this, we find out that

$$E(\mathbf{F}_5) = \{(0,1), (0,4), (2,3), (2,2), (4,0)\}.$$

EXERCISE: List the points in $E(\mathbf{F}_7)$.

There are two lies in what we've told you above, and both of them will get us into trouble before long if we do not correct them now. First, an elliptic curve can only be written using the equation $y^2 = x^3 + Ax + B$ as long as the number system R we use does not contain $\mathbf{F}_2$ or $\mathbf{F}_3$. This is not such a big lie. There are more complicated equations that allow us to define an elliptic curve that will work in all number systems.[3]

The second lie is a bigger one. In addition to the solutions to the cubic $\mathbf{Z}$-equation, an elliptic curve contains an additional point, which is written $\mathcal{O}$ and is called "the point at infinity." Roughly speaking, it corresponds to the solution $x = \infty$, $y = \infty$, but we will

[3]We will hide it in a footnote, but here is the equation:

$$y^2 + a_1xy + a_3y = x^3 + a_2x^2 + a_4x + a_6,$$

along with a very complicated expression involving a_1, a_2, a_3, a_4, and a_6 that is not allowed to be 0. Strictly speaking, we have to be careful whenever $\frac{1}{6}$ is not in R, so even $E(\mathbf{Z})$ must be taken with a grain of salt.

rarely write it like that. So the correct answer is that

$$E(\mathbf{F}_5) = \{(0,1),(0,4),(2,3),(2,2),(4,0),\mathcal{O}\},$$

and

$$\begin{aligned} E(\mathbf{F}_7) &= \{(0,1),(0,6),(1,3),(1,4),(2,3),(2,4),(3,0),\\ &\quad (4,3),(4,4),(5,0),(6,0),\mathcal{O}\}. \end{aligned}$$

What is the reason to add this extra solution? We need it in order to make the set of solutions to the **Z**-equation into a group. In fact, $\mathcal{O}$ is the neutral element for the group law.

The Group Law on an Elliptic Curve

What is the definition of the group operation? It is pretty complicated, because it has a lot of parts. Let $P = (x_1, y_1)$ and $Q = (x_2, y_2)$ be two points in $E(R)$, for some field R. We are going to tell you how to combine them. Even though what we are going to do is much more complicated algorithmically than anything we have done so far, mathematicians still just use the symbol "+" to stand for the group law on an elliptic curve. We use this symbol even though addition of integers is about the simplest group law that there is and combining points on an elliptic curve may be the most complicated group law that you will ever encounter. The symbol "+" reminds us that the commutative law holds when we combine points on an elliptic curve.

This is a group law that is easily worked on a computer. What is the rule for combining two points P and Q on an elliptic curve E to get another point on that same elliptic curve?

1. If $P = \mathcal{O}$, then $P + Q = Q$.
2. If $Q = \mathcal{O}$, then $P + Q = P$.
3. If $x_1 = x_2$ and $y_1 + y_2 = 0$, then $P + Q = \mathcal{O}$. Here, $y_1 + y_2 = 0$ means equality in the field R, and similarly in the remainder of our description.
4. If $x_1 \neq x_2$, then compute the numbers $\lambda = \frac{y_2 - y_1}{x_2 - x_1}$ and $\nu = y_1 - \lambda x_1$. (As usual, the arithmetic operations take

place according to the rules of the field R.) Let $x_3 = \lambda^2 - x_1 - x_2$. Let $y_3 = -(\lambda x_3 + \nu)$. Then $P + Q$ has coordinates (x_3, y_3).

5. If $x_1 = x_2$ and $y_1 = y_2 \neq 0$, then compute the number $\lambda = \frac{3x_1^2+A}{2y_1}$. Compute ν as before, and then use the same formula for x_3 and y_3 as in the previous case.

A Much-Needed Example

This description is somewhat mysterious.[4] In fact, it is not obvious (unless you looked at the footnote) that (x_3, y_3) even solves the same equation $y^2 = x^3 + Ax + B$ that we started with. We take some examples in $E(\mathbf{Q})$ with the particular curve $y^2 = x^3 + 1$, which means that $A = 0$ and $B = 1$. Remember that $(0, 1)$, $(0, -1)$, and $(2, 3)$ are points on this curve. The first case of the group law tells us that $\mathcal{O} + (0, 1) = (0, 1)$, and the second case tells us that $(0, 1) + \mathcal{O} = (0, 1)$. The third case tells us that $(0, 1) + (0, -1) = \mathcal{O}$.

Now things get messier. We compute $(0, 1) + (2, 3)$. We start by computing $\lambda = \frac{3-1}{2-0} = \frac{2}{2} = 1$. Next, $\nu = 1 - 1 \cdot 0 = 1$. Then $x_3 = 1^2 - 0 - 2 = -1$, and $y_3 = -(1 \cdot -1 + 1) = 0$. So we just worked out that $(0, 1) + (2, 3) = (0, -1)$.

Finally, we compute $(2, 3) + (2, 3)$. Using the last case, we compute that $\lambda = \frac{3 \cdot 2^2 + 0}{2 \cdot 3} = \frac{12}{6} = 2$. Next, $\nu = 3 - 2 \cdot 2 = -1$. Now, $x_3 = 4 - 2 - 2 = 0$ and $y_3 = -(2 \cdot 0 + (-1)) = 1$. So $(2, 3) + (2, 3) = (0, 1)$.

EXERCISE: Let E be the elliptic curve $y^2 = x^3 + 17$. Compute $(-1, 4) + (2, 5)$ and $(2, 5) + (2, 5)$.

SOLUTION: To compute $(-1, 4) + (2, 5)$, we first compute that $\lambda = \frac{1}{3}$, and then $\nu = \frac{13}{3}$, and then $(-1, 4) + (2, 5) = (-\frac{8}{9}, -\frac{109}{27})$.

To compute $(2, 5) + (2, 5)$, we first compute that $\lambda = \frac{6}{5}$, and then $\nu = \frac{13}{5}$, and then $(2, 5) + (2, 5) = (-\frac{64}{25}, \frac{59}{125})$.

---

[4]Geometrically, what is going on is this: We take P and Q and connect them with a line. This line will intersect the elliptic curve in exactly three points: P, Q, and a third point T. We negate the y coordinate of T and the resulting point is $P + Q$. This is what underlies formula (4). The other formulas have similar geometric interpretations. The fact that this geometric construction defines a group law is amazing and not so easy to prove.

The particular details of the abelian group law actually do not concern us very much. What does matter is that if we combine two points in $E(\mathbf{Q})$, for example, then the result is also in $E(\mathbf{Q})$. This is not true for $E(\mathbf{Z})$ because the number λ might be a fraction, and then we might get fractional values for x_3 and y_3. But it is true for any number system in which we can divide. So if we start with elements of $E(\mathbf{R})$, and add them, we will get elements of $E(\mathbf{R})$; ditto for $E(\mathbf{C})$ and $E(\mathbf{F}_p)$.

Digression: What Is So Great about Elliptic Curves?

The question should be, "What is *not* great about them?" They were essential in Wiles's proof of Fermat's Last Theorem. You can read a brief description in our paper (Ash and Gross, 2000), or look at some of the many books on the subject (Hellegouarch, 2002; Singh, 1997; van der Poorten, 1996), or skip ahead to chapter 22. Elliptic curves have given rise to a tremendous amount of interesting number theory, including the part we are trying to explain in this book. Mathematicians are just lucky that elliptic curves are fairly simple—they only involve two variables and no powers higher than the cube—and yet are so rich.

One can see how people began to be curious about elliptic curves—even ancient Greeks such as Diophantus, although he did not call them elliptic curves. As we have seen, equations in only one variable can be difficult to solve; even formulas for the cubic and quartic equations were not discovered until the Renaissance. But we go on to one equation in two variables anyway. If all the exponents are 1 or less, the solution set can be found by simple arithmetic. Uninteresting. (See chapter 10 about how you can use matrices to solve such equations.) If all the exponents are 2 or less, the solution set can be understood in terms of what are called "conic sections," and some of the later Greeks understood those very well, although it was a fairly advanced topic for them. So the natural thing to do, if you are a mathematician, is to keep going until you get stuck: Throw in a cubic exponent. Just one for starters and, boom, you get stuck right away. *Voilà*—elliptic curves.

The Congruent Number Problem

Another interesting Greek problem that turns out to be tied up with elliptic curves is what might be called a Diophantine–Pythagorean problem: Find all right triangles whose side lengths are rational numbers and whose area is the integer D. This is also called the "congruent number problem." It does not sound so difficult, does it? But it *is* difficult, and it has not yet been fully solved.

If you want to do a little algebra, you can see how elliptic curves come up. We look at the case where $D = 1$. Using only rational numbers, we want to solve the system of equations:

1. $x^2 + y^2 = z^2$.
2. $xy/2 = 1$.

These equations correspond to a right triangle whose sides have lengths $|x|$ and $|y|$, with hypotenuse $|z|$. Divide equation (1) through by z and set $X = x/z$, $Y = y/z$. Then our system is equivalent to

1′. $X^2 + Y^2 = 1$.
2′. $XY/2 = 1/z^2$.

On page 54 in the subsection on **Z**-equations, we pointed out that if we set $w = 1 + t^2$, the solutions to (1′) are all of the form $X = (1 - t^2)/w$ and $Y = 2t/w$, where t can take the value of any rational number. (Oh yes, there is also the solution $X = -1$, $Y = 0$.)

Plugging this into (2′), multiplying through by w^2, and cancelling the 2's, we get

3. $t - t^3 = (w/z)^2$.

You may think this looks bad, but remember that we can make z anything we need it to be to make item (3) hold. But we do have to be able to take the square root of the left side of (3) in order to solve for z. In other words, the Diophantine–Pythagorean problem we started with is equivalent to finding all rational numbers t such that $t - t^3$ is a square of a rational number. That is, we want to solve for the solution set $E(\mathbf{Q})$ where E is the elliptic curve given by

$$y^2 = -t^3 + t.$$[5]

By the way, the "extra" solution $X = -1$, $Y = 0$ to $X^2 + Y^2 = 1$ corresponds to the point $\mathcal{O}$ at infinity on this elliptic curve.[6] However, it is not much of a triangle: One of the sides is 0. In fact, we could have excluded the solutions where X or Y is 0 to begin with. But we would not have seen how the point at infinity on the elliptic curve can correspond to a "degenerate solution" to our original problem: a triangle with one side 0, the other side ∞, in such a way that the area is 1.

But the real question here is whether there are any nondegenerate rational solutions in $E(\mathbf{Q})$. In a course in elementary number theory, you can prove that the answer is "no." So when $D = 1$, the congruent number problem has a negative solution: There is no right triangle with rational sides and area 1. For more on the congruent number problem, see chapter 23.

Torsion and the Galois Group

It is very difficult to prove that the five-step recipe that we have given above actually defines a group law. It is not obvious that it satisfies the associative law. We do have obvious inverses: the inverse of (x_1, y_1) is $(x_1, -y_1)$. And we do have an identity element: $\mathcal{O}$. But you will have to take associativity on faith; the algebra needed to verify it is rather involved.[7] (Commutativity of the group law is not difficult to check from the definition.)

The group law is a big help when studying $E(\mathbf{Q})$. It can help you derive new solutions from old ones, and sometimes can even help

---

[5]You may have noticed that this does not conform to our template $y^2 = x^3 + Ax + B$ for elliptic curves. To fix this, define a new variable $u = -t$, and get the equivalent equation $y^2 = u^3 - u$.

[6]Why? Well, we get this extra point if we set $u = \infty$, as it were. That would correspond to the point (∞, ∞) on the elliptic curve.

[7]Associativity can also be proved in more elegant ways using algebraic geometry.

to prove that there are no unobvious solutions—all depending on which curve E you are studying.

There are many theorems about the set $E(\mathbf{Z})$ and the group $E(\mathbf{Q})$ that can be found in the textbooks. For our purposes, in studying representations of the absolute Galois group G of $\mathbf{Q}$, we need to tell you about a particular set of elements inside of $E(\mathbf{C})$: the *n-torsion*.

DEFINITION: Let n be a positive integer. An element P of $E(\mathbf{C})$ is part of the *n-torsion* if

$$\overbrace{P+P+\cdots+P}^{n \text{ times}} = \mathcal{O}.$$

The set of all n-torsion points is written $E[n]$.

For example, an element P on the elliptic curve is part of the 5-torsion if $P+P+P+P+P=\mathcal{O}$. This is usually written, rather confusingly, as $5P=\mathcal{O}$. Do not confuse it with multiplying the x- and y-coordinates of P by 5, because it means something much more complicated. You can work out what it does mean: It will give some $\mathbf{Z}$-equation that the x-coordinate of P must satisfy, and another one for the y-coordinate of P, for P to be 5-torsion. Therefore, the x- and y-coordinates of P will be in $\mathbf{Q}^{\text{alg}}$.

In fact, the equation that the x-coordinate must solve will be a rather messy degree-12 $\mathbf{Z}$-equation, and so there will be no more than 25 elements of the 5-torsion (for any particular elliptic curve, of course; if you change the elliptic curve, then you will get a different equation, and different solutions).[8] The general situation is summarized as follows.

THEOREM 9.1: Let E be an elliptic curve, and let n be a positive integer. All of the n-torsion points of E have coordinates in $\mathbf{Q}^{\text{alg}}$, and the number of elements of $E(\mathbf{Q}^{\text{alg}})$ that are n-torsion is n^2.

---

[8]How do we get 25 elements? The equation for the x-coordinate will be degree 12, whose roots will give us at most 12 x-coordinates. Each of those corresponds to 2 y-coordinates, which are $\pm$ of each other. That is 24 points. The remaining point is $\mathcal{O}$.

This gives us a chance to play all of the games with the Galois group that we can play with any **Z**-equation. In particular, we can take the n-torsion—we call the points $P_1, \ldots, P_{n^2}$—and pick some element g in the Galois group G, and compute $g(P_1)$. By this, we mean compute g of the x-coordinate and g of the y-coordinate. Because these coordinates solve **Z**-polynomials when we apply g, we must get some other solution of the same polynomials, meaning some other point in the n-torsion. So $g(P_1)$ must be P_1, or P_2, or P_3, and so on. This means that we can learn about our element g by thinking of it as a permutation of the elements of the n-torsion for any n and for any elliptic curve E.[9]

We will return to the torsion of elliptic curves in chapter 18 on one- and two-dimensional representations of the Galois group.

You may be wondering about the funny word "torsion." Its root meaning is "twist." We call any element in a group that satisfies the equation $\overbrace{g \circ g \circ \cdots \circ g}^{n \text{ times}} = e$ an *n-torsion element*. The term seems to have arisen in topology, where torsion elements in certain groups described "twisting" in certain topological spaces. When the group is an elliptic curve, the equation $g^n = e$ is written as $nP = \mathcal{O}$.

[9]In case you are worried, we define $g(\mathcal{O})$ to be $\mathcal{O}$.

Chapter 10

MATRICES

Road Map

We review matrices from the ground up, because they will play a major role in our story from now on. Our main focus is matrix multiplication, which will be used as a group operation in the next chapter, thereby giving us a large supply of groups of matrices which we understand quite well. We will then use matrix groups as standard objects—targets—in Galois representation theory. (This could be a good time to review chapter 1.)

Matrices and Matrix Representations

We need matrices in order to get to the center of our subject—representations of Galois groups. There are two types of representations we must consider: matrix representations and permutation representations. We will discuss matrix representations in chapter 12 and permutation representations in chapter 14.

The two kinds of representations are closely related. Because we defined the Galois group as a group of permutations, one may think that it would be enough to study permutation representations. But even if we knew everything about the permutation representations, we would need the matrix kind in order to formulate some of the deeper properties of Galois groups and **Z**-varieties. We shall see how this surprising fact plays out in the final chapters of this book.

Matrices are important, even if we have no interest in number theory, let alone Galois groups. They belong to algebra, but they are also important in topology and geometry, not to mention physics, chemistry, economics, and many other areas. For example, the study of special relativity can be viewed as the theory of a certain group of 4-by-4 matrices, the *Lorentz group*. Elementary quantum mechanics involves representations of operators by matrices, some of them of infinite size. Economists use matrices in game theory. And so on.

There are groups of matrices that provide us with a series of standard examples of groups whose group law is not commutative (i.e., $x \circ y$ is not necessarily equal to $y \circ x$). This partially explains why they are so useful. Many of our most important groups, including SO(3) and the absolute Galois group G, have a noncommutative group law. If we want to get a useful snapshot of such groups by representing them inside a standard group, we need a large supply of noncommutative standard groups. We get this supply from the permutation and matrix groups.

A word about terminology: It gets boring using the word *matrix* both as a noun, naming the entity we are soon to define, and as an adjective. So we have a synonym for the adjectival use: *linear*. A *linear representation* is just a matrix representation, and a *linear group* is just a group of matrices. BEWARE: The adjective "linear" has many other meanings, such as "in the shape of a straight line," and it has other uses in mathematics and even in representation theory. We hope these multiple meanings will not be a problem in the explanations to come. We will usually use linear to mean matricial.

Matrices and Their Entries

What is a matrix? In mathematics, the usual definition is "a rectangular array of objects." This is a correct definition, but it seems too vague to be of much use. It conjures up armies of objects marching past, and perhaps forming and reforming their arrays. We try to understand this definition as concretely as possible.

First, we discuss the objects that are going to appear in the matrices. We do not really want to consider just any old objects.

It is true that technically this is a matrix:

$$\begin{bmatrix} \pounds & \aleph \\ \heartsuit & \Delta \end{bmatrix}$$

But we do not want this kind of thing sneaking up into a theory, any more than a platoon of soldiers. So we agree that all the objects in our rectangular array will be elements of some number system R. We call such a matrix an R-matrix. For example, we refer to a $\mathbf{Z}$-matrix, or a $\mathbf{C}$-matrix, or an $\mathbf{F}_2$-matrix.

We call our matrix an r-by-c matrix if it has r rows and c columns. Consider, for example, the particular number system $\mathbf{F}_2$, consisting of the numbers 0 and 1. Some $\mathbf{F}_2$-matrices are the following:

$$\begin{bmatrix} 1 \end{bmatrix}, \quad \begin{bmatrix} 1 & 0 \end{bmatrix}, \quad \begin{bmatrix} 1 & 1 & 0 \end{bmatrix}, \quad \begin{bmatrix} 0 \\ 0 \end{bmatrix}, \quad \begin{bmatrix} 1 \\ 0 \\ 1 \end{bmatrix},$$

$$\begin{bmatrix} 1 & 1 & 0 & 1 \\ 1 & 0 & 1 & 1 \end{bmatrix}$$

EXERCISE: Draw all possible 2-by-3 $\mathbf{F}_2$-matrices. How many are there? (You may need a fairly large sheet of paper for this.)

If we want to specify a matrix, we can draw it or we can say which numbers are where. For example, the $\mathbf{F}_2$ matrix $\begin{bmatrix} 1 & 0 & 0 \\ 0 & 1 & 0 \end{bmatrix}$ can be drawn just like that, or we can tell you that the first row is $[1 \quad 0 \quad 0]$ and the second row $[0 \quad 1 \quad 0]$, or we can tell you that the first column is $\begin{bmatrix} 1 \\ 0 \end{bmatrix}$, the second is $\begin{bmatrix} 0 \\ 1 \end{bmatrix}$, and the third is $\begin{bmatrix} 0 \\ 0 \end{bmatrix}$. Or, most painstaking of all, we can phone you the following instructions: Put a 1 in the first row and first column, put a 0 in the first row and second column, and so on. Although this last method is painful, it does not matter if we get the order of the instructions mixed up. It also tells you how you can inquire about a specific element: "There was static on the line and I didn't hear what number was in the second row and first column. Oh, it was a 0? . . ."

We call the actual numbers that occur in a given matrix its *entries*. So the last exchange can be stated: "What was the entry in the second row and first column? Oh, it was a 0?" Or when you get good at it, you can even say "What was the (2,1)-entry?"

Matrix Multiplication

Because the entries in a matrix are elements of a number system, they can be added, subtracted, and multiplied. This allows us to define multiplication of matrices, and thus we will obtain plenty of "standard" groups to use in our representations.

Here is the all-important definition of how to multiply two matrices. First, we have to tell you that any two matrices cannot necessarily be multiplied. They must both have entries from the same number system. For instance, both could be **Z**-matrices, or **Q**-matrices, or **C**-matrices.

The sizes of the matrices also have to match in a particular way. Remember that an m-by-n matrix is a rectangular array of numbers with m rows and n columns. Let A be an m-by-n R-matrix and B a p-by-q R-matrix. Then we can form and compute the product AB (in that order!) if and only if $n = p$. In words: The number of columns of the first matrix must be the same as the number of rows of the second matrix. You will see why as soon as we explain how to multiply.[1] A mnemonic for this rule is that the numbers in the middle have to "cancel out." In fact, an m-by-n matrix times an n-by-q matrix will be an m-by-q matrix, the n's having cancelled out.

It will be helpful first to describe the dot product of a row and column of the same sizes. We assume the number system is fixed for this discussion, so that we can add and multiply entries. The dot product of a row of length n and a column of length n is obtained by multiplying the pairs of entries in corresponding positions in

[1]You may have noticed that we do not use any symbol for the group law of multiplying matrices. We just juxtapose the symbols of the two matrices, as in AB. This notation is traditional.

the row and column, and then adding up all of those answers. For example:

The dot product of $\begin{bmatrix} 1 & 3 & 5 \end{bmatrix}$ and $\begin{bmatrix} 2 \\ 9 \\ 1 \end{bmatrix}$ is $1 \cdot 2 + 3 \cdot 9 + 5 \cdot 1 =$

$2 + 27 + 5 = 34.$

So the dot product of a row and a column (of the same size) is a single number.

Next, let's multiply a 2-by-2 **Z**-matrix A by a 2-by-1 **Z**-matrix B:

$$\begin{bmatrix} 2 & 5 \\ 4 & 1 \end{bmatrix} \begin{bmatrix} 6 \\ 8 \end{bmatrix} = \begin{bmatrix} 2 \cdot 6 + 5 \cdot 8 \\ 4 \cdot 6 + 1 \cdot 8 \end{bmatrix} = \begin{bmatrix} 52 \\ 32 \end{bmatrix}.$$

The answer is another 2-by-1 **Z**-matrix. The multiplication rule can be stated in words as follows: The first entry in the answer is the dot product of the first row of A by B, and the second entry in the answer is the dot product of the second row of A by B.

Or here's a 4-by-3 **Z**-matrix times a 3-by-1 **Z**-matrix:

$$\begin{bmatrix} 1 & 2 & 3 \\ 7 & 8 & 1 \\ -1 & 2 & 3 \\ 1 & 11 & 0 \end{bmatrix} \begin{bmatrix} 40 \\ 28 \\ 31 \end{bmatrix} = \begin{bmatrix} 1 \cdot 40 + 2 \cdot 28 + 3 \cdot 31 \\ 7 \cdot 40 + 8 \cdot 28 + 1 \cdot 31 \\ -1 \cdot 40 + 2 \cdot 28 + 3 \cdot 31 \\ 1 \cdot 40 + 11 \cdot 28 + 0 \cdot 31 \end{bmatrix} = \begin{bmatrix} 189 \\ 535 \\ 109 \\ 348 \end{bmatrix}.$$

Again, each entry in the answer is a dot product.

In general, an m-by-n matrix times an n-by-1 matrix will be an m-by-1 matrix computed according to this rule: the kth entry of the answer is the dot product of the kth row of the first matrix times the column matrix.

EXERCISE: Compute the following products:

$$\begin{bmatrix} 5 & 6 \\ 3 & 4 \\ 11 & -2 \end{bmatrix} \begin{bmatrix} 4 \\ 9 \end{bmatrix} \qquad \begin{bmatrix} 4 & 3 & 4 \\ 1 & 8 & 2 \end{bmatrix} \begin{bmatrix} 7 \\ 9 \\ 12 \end{bmatrix}.$$

SOLUTION: The answers are

$$\begin{bmatrix} 5 & 6 \\ 3 & 4 \\ 11 & -2 \end{bmatrix} \begin{bmatrix} 4 \\ 9 \end{bmatrix} = \begin{bmatrix} 74 \\ 48 \\ 26 \end{bmatrix}$$

and

$$\begin{bmatrix} 4 & 3 & 4 \\ 1 & 8 & 2 \end{bmatrix} \begin{bmatrix} 7 \\ 9 \\ 12 \end{bmatrix} = \begin{bmatrix} 103 \\ 103 \end{bmatrix}.$$

Now multiplying a matrix by another matrix (of the right size) is easy. An m-by-n matrix times an n-by-q matrix will be an m-by-q matrix computed according to this rule: the (k,j)th entry of the answer is the dot product of the kth row of the first matrix and the jth column of the second matrix.

That was a mouthful, so let us take a look. We have

$$\begin{bmatrix} 1 & 2 \end{bmatrix} \begin{bmatrix} 3 & 4 \\ 5 & 6 \end{bmatrix} = \begin{bmatrix} 1 \cdot 3 + 2 \cdot 5 & 1 \cdot 4 + 2 \cdot 6 \end{bmatrix} = \begin{bmatrix} 13 & 16 \end{bmatrix}.$$

Here is another example:

$$\begin{bmatrix} 3 & 4 & 11 \\ 2 & 4 & -1 \end{bmatrix} \begin{bmatrix} 4 & 5 & 6 & 11 \\ 2 & 2 & 1 & 34 \\ 9 & 4 & 32 & 37 \end{bmatrix} =$$

$$\begin{bmatrix} 3 \cdot 4 + 4 \cdot 2 + 11 \cdot 9 & 3 \cdot 5 + 4 \cdot 2 + 11 \cdot 4 & 3 \cdot 6 + 4 \cdot 1 + 11 \cdot 32 & 3 \cdot 11 + 4 \cdot 34 + 11 \cdot 37 \\ 2 \cdot 4 + 4 \cdot 2 + -1 \cdot 9 & 2 \cdot 5 + 4 \cdot 2 + -1 \cdot 4 & 2 \cdot 6 + 4 \cdot 1 + -1 \cdot 32 & 2 \cdot 11 + 4 \cdot 34 + -1 \cdot 37 \end{bmatrix} =$$

$$\begin{bmatrix} 119 & 67 & 374 & 576 \\ 7 & 14 & -16 & 121 \end{bmatrix}.$$

EXERCISE: Compute the following matrix products:

$$\begin{bmatrix} 7 & 8 & 11 \\ 4 & 3 & 12 \end{bmatrix} \begin{bmatrix} 8 & 5 \\ 3 & 9 \\ 11 & 2 \end{bmatrix} \qquad \begin{bmatrix} 8 & 5 \\ 3 & 9 \\ 11 & 2 \end{bmatrix} \begin{bmatrix} 7 & 8 & 11 \\ 4 & 3 & 12 \end{bmatrix}.$$

SOLUTION: We have

$$\begin{bmatrix} 7 & 8 & 11 \\ 4 & 3 & 12 \end{bmatrix} \begin{bmatrix} 8 & 5 \\ 3 & 9 \\ 11 & 2 \end{bmatrix} = \begin{bmatrix} 201 & 129 \\ 173 & 71 \end{bmatrix}$$

and

$$\begin{bmatrix} 8 & 5 \\ 3 & 9 \\ 11 & 2 \end{bmatrix} \begin{bmatrix} 7 & 8 & 11 \\ 4 & 3 & 12 \end{bmatrix} = \begin{bmatrix} 76 & 79 & 148 \\ 57 & 51 & 141 \\ 85 & 94 & 145 \end{bmatrix}.$$

The process by which we have defined matrix multiplication is typically mathematical. First we define and understand a special case: a row times a column, that is, a 1-by-n matrix times an n-by-1 matrix, will be a 1-by-1 matrix—a single number—given by the dot product. From this we build up: next defining any matrix times a column, and finally any matrix times any matrix (as long as the sizes match).

Linear Algebra

One simple use of matrices is to solve systems of **Z**-equations where none of the exponents is higher than 1. This is called "linear algebra."[2] Here is an example.

Suppose we want to solve the system of **Z**-equations:

$$3x - 5y = 2. \tag{10.1}$$

$$2x + 3y = 14. \tag{10.2}$$

This is easy to solve using elementary algebra. And it is also easy to graph the two lines (which is why it is called *linear* algebra) and see where they intersect. But in order to illustrate matrices, let's set this up as a matrix problem.

[2]This is one reason why "linear" is often used as an adjective in place of "matricial."

We define the **Z**-matrix A to be

$$A = \begin{bmatrix} 3 & -5 \\ 2 & 3 \end{bmatrix}$$

and we define another **Z**-matrix B to be

$$B = \begin{bmatrix} 2 \\ 14 \end{bmatrix}.$$

Finally, we define an unknown matrix Z in terms of x and y:

$$Z = \begin{bmatrix} x \\ y \end{bmatrix}.$$

The value Z is just as unknown as the x and y we are trying to find.

Now by practicing your matrix multiplication, you can see that the system given by (10.1) and (10.2) above is completely equivalent to the single matrix equation:

$$AZ = B.$$

It turns out there is a way to divide by the matrix A and solve for Z. We will discuss this in more detail in the next chapter. For now, we just define two more matrices:

$$A^{-1} = \begin{bmatrix} \frac{3}{19} & \frac{5}{19} \\ \frac{-2}{19} & \frac{3}{19} \end{bmatrix}$$

and

$$I = \begin{bmatrix} 1 & 0 \\ 0 & 1 \end{bmatrix}.$$

Check that $A^{-1}A = I$ and $IZ = Z$.

If $AZ = B$ then $A^{-1}(AZ) = A^{-1}B$. In the next chapter we will find out that matrix multiplication is associative, so that $A^{-1}(AZ) = (A^{-1}A)Z = IZ = Z$. We conclude that if $AZ = B$, then

$$Z = A^{-1}B = \begin{bmatrix} \frac{3}{19} & \frac{5}{19} \\ \frac{-2}{19} & \frac{3}{19} \end{bmatrix} \begin{bmatrix} 2 \\ 14 \end{bmatrix} = \begin{bmatrix} \frac{76}{19} \\ \frac{38}{19} \end{bmatrix} = \begin{bmatrix} 4 \\ 2 \end{bmatrix}.$$

Thus $Z = \begin{bmatrix} 4 \\ 2 \end{bmatrix}$, and indeed you can check that $x = 4$, $y = 2$ solves the original system (10.1)–(10.2).

Just by glancing at them, who would have thought that the number 19 was lurking in the system of equations (10.1)–(10.2)? But if you change the constants on the right-hand side of the equations, you are likely to find 19's popping up in the denominators of the solutions. This is related to the fact that the determinant of the matrix A is 19 (see chapter 11).

There is a complete theory of this and it gives you very efficient ways to solve many simultaneous equations in many unknowns, as long as none of the exponents is greater than 1. If you have many equations and unknowns, you should probably use a computer to help you, but the computer will be using matrices. At the end of the next chapter we will do a 3-by-3 example.

Digression: Graeco-Latin Squares

Matrices with entries that are not numbers can still have uses. For example, a game from the 1950s consisted of a 6-by-6 grid of squares and 36 pieces. The plastic pieces were of six different colors and six different shapes, so that *each piece was different from all the others*. The challenge was to arrange the pieces on the grid in such a way that no row or column contained two pieces of the same shape or same color. The manufacturer offered a prize of $1000 for the first solution. Some people tried this for quite a while.

In fact, the eighteenth-century mathematician Leonhard Euler had already publicized this problem, which has a long subsequent history. Instead of plastic pieces, he worked with soldiers, each of which had a rank and a regiment. It is customary nowadays to think of a square matrix, say with n rows and n columns. The set of entries consists of n^2 symbols of the form $A\alpha$, that is, one Roman letter chosen from a list of n Roman letters, and one Greek letter chosen from a list of n Greek letters, in such a way that the entries are all different from one another. (If n is greater than 24, find lengthier alphabets.)

DEFINITION: A matrix of these n^2 entries satisfying the condition that no row and no column contains the same Latin or the same Greek letter twice is called a *Graeco-Latin Square*.

For example,

$$\begin{bmatrix} A\alpha & B\beta & C\gamma & D\delta \\ D\gamma & C\delta & B\alpha & A\beta \\ C\beta & D\alpha & A\delta & B\gamma \\ B\delta & A\gamma & D\beta & C\alpha \end{bmatrix}$$

is a 4-by-4 Graeco-Latin Square. It is easy to see that there is no 2-by-2 Graeco-Latin Square.

These squares are actually useful in experimental design, when you want to test all possible combinations of two variables. By the way, there is no 6-by-6 Graeco-Latin Square. Alas. (This fact was proven by a very patient mathematician in 1901.)

In fact, after mathematicians showed that there is no 2-by-2 Graeco-Latin Square, and no 6-by-6 Graeco-Latin Square, it was tempting to guess (as Euler did) that there was no 10-by-10 Graeco-Latin Square. However, this turned out to be wrong (though you are unlikely to find one just by using trial-and-error without a computer).[3]

[3]There *are* 8-by-8 Graeco-Latin Squares. In fact, Euler knew that there were n-by-n Graeco-Latin Squares for any odd n or n divisible by 4. Euler conjectured that all $4k+2$-by-$4k+2$ cases (for $k \geq 0$) were impossible, but in 1959 it was proven that all of them are possible, *except* for the 2-by-2 and 6-by-6 cases!

Chapter 11

GROUPS OF MATRICES

Road Map

We continue to study matrices. Some of them are "invertible," which means they can belong to multiplicative groups of matrices. These matrix groups are the standard objects we referred in the road map of the previous chapter.

In particular, for any number system R, we define the group $\mathrm{GL}(n,R)$ of all invertible R-matrices with n rows and n columns. As an example, we explore $\mathrm{GL}(2,\mathbf{Z})$, an infinite group of special importance in number theory.

Square Matrices

From now on, we will assume that the set of entries in our matrices consists of elements of a number system, which we will call R. This means that we can add, subtract, and multiply the elements of R, and they obey the usual laws of arithmetic. We will not need to use division at all, and so letting $R = \mathbf{Z}$, for example, is fine.

The entries need not be actual numbers of the ordinary sort. For example, we could use the field $\mathbf{F}_2$ consisting of 0 and 1. Let us review a part of chapter 4. We define addition as usual, and multiplication as usual, with one important exception: $1 + 1 = 0$. (Because there is no "2" in $\mathbf{F}_2$, we have to redefine $1 + 1$. When someone tells you that a given fact is as obvious as $1 + 1 = 2$, tell her that maybe she should go visit $\mathbf{F}_2$.) You can figure out how to redefine subtraction also.[1] Then all the laws of arithmetic hold: associative laws of addition and multiplication, commutative laws of addition and multiplication, distributive laws, adding 0 does not change the number you add it to, and neither does multiplying by 1. Later in this chapter, we will study the group $GL(2, \mathbf{F}_2)$.

We explore the following question: When is a set of R-matrices a group, with the group law given by matrix multiplication? Remember that this means (see chapter 2):

1. The set is "closed" under the group operation.[2]
2. The associative law holds.
3. There is a neutral element.
4. Any element in the set has an inverse element in the set.

The first thing to verify is the nonobvious fact that matrix multiplication is always associative. You can find the proof in any linear algebra textbook. Even if you do not want to prove it, you can check an example or two for yourself. Pick three $\mathbf{Z}$-matrices, A, B, and C of matching sizes: the number of columns of A = the number of rows of B, and the number of columns of B = the number of rows of C. Then compute $(AB)C$—multiply AB first and then multiply the result by C on the right. Next compute $A(BC)$—multiply BC first and then multiply the result by A on the left. If you did not make any errors, the answers should be the same.

By now, you may have noticed that the set of all R-matrices is not closed under matrix multiplication for the unusual reason that you cannot even multiply two matrices if their sizes do not match. A group law has to be able to take any two elements of the group and combine them to get a third element of the group. If we try to limit our set of matrices to some set where every matrix matches every other one in the set for multiplicative purposes, we see that we have to limit ourselves to *square* matrices of some fixed size.

So fix a positive integer n, and now let us consider only n-by-n R-matrices. There is indeed a neutral element for matrix

---

[1]The only deviation from usual subtraction is that $0 - 1 = 1$.

[2]DEFINITION: A set S is *closed under a group law* if $x \circ y$ is in S whenever both x and y are in S.

multiplication in this set: It is called the *identity matrix*. It has 1's down the main diagonal that goes from upper left to lower right (this is simply called "*the* diagonal" when dealing with matrices) and it has 0's everywhere else. For instance, here is the 3-by-3 identity matrix:

$$\begin{bmatrix} 1 & 0 & 0 \\ 0 & 1 & 0 \\ 0 & 0 & 1 \end{bmatrix}.$$

It is traditional to call this matrix I. Of course, if you change n, you change I, so it should really be called I_n, but the size of I will be clear from the context.

EXERCISE: Take this 3-by-3 matrix I and multiply it by any 3-by-3 matrix A. Try it on the left and the right: AI and IA both will turn out to be A again.

Matrix Inverses

Great! All we need are inverses. But now things get a little sticky, or, as a mathematician would say, interesting. Not every matrix has an inverse.

DEFINITION: An n-by-n matrix A is *invertible*, or *has an inverse*, if there is some n-by-n matrix H with the property that AH and HA both equal I, the n-by-n identity matrix. If this is so, we say that H is the *inverse* of A.

Here is a 3-by-3 example of an invertible matrix A:

$$\begin{bmatrix} 2 & 3 & 4 \\ 1 & 0 & 2 \\ 3 & 4 & 8 \end{bmatrix}.$$

And here is its inverse H:

$$\begin{bmatrix} \frac{4}{3} & \frac{4}{3} & -1 \\ \frac{1}{3} & -\frac{2}{3} & 0 \\ -\frac{2}{3} & -\frac{1}{6} & \frac{1}{2} \end{bmatrix}.$$

EXERCISE: Check that $AH = HA = I$.

It is a fact, easy to prove, that if A is invertible, then it has one and only one inverse.[3] This we call *the* inverse of A, and we write it A^{-1}. Notice this doesn't mean $\frac{1}{A}$, for A is not a number, so it cannot be divided into the number 1.

We claimed that not every square matrix has an inverse. Here is a stupid example. If C is the 3-by-3 matrix of all 0's, then $CB = C$ no matter what 3-by-3 matrix B is. So CB could never equal I. But there are less stupid examples. If you think about it a little, you can see that even if just one row of A is all 0's, then A has no inverse matrix. Or if two of the rows of A are identical, then A has no inverse matrix. Or if two of the columns of A are identical, then A has no inverse matrix. Do you see why?[4]

The general theory of which A's are invertible and which are not is a very interesting part of matrix theory. The complete answer is too much off the track for us to go into it here, but if you happen to know how to compute the determinant of a square matrix, then you can understand the following:

THEOREM 11.1: If A is a square R-matrix, then A is invertible if and only its determinant (which is an element of R) has a multiplicative inverse in R. That is to say, if d is the determinant of A, then A is invertible if and only if there is some element c in R so that the product $cd = 1$.

---

[3] PROOF: If H and K are both inverses of A, then $H = HI = H(AK) = (HA)K = IK = K$.

[4] For example, if A has two identical rows, then so does AB for any matrix B. But the identity matrix does not have two identical rows. So AB can never equal I, whatever matrix B you try.

A nice example in which we can write everything down is given by 2-by-2 $\mathbf{F}_2$ matrices. There are 16 of them (two choices for each of the four entries). An $\mathbf{F}_2$-matrix is invertible unless it has an all 0 row, an all 0 column, or two identical rows. After throwing out all of these, we find that the set of *invertible* 2-by-2 $\mathbf{F}_2$ matrices has exactly six elements. For instance:

EXERCISE: Why isn't the $\mathbf{F}_2$-matrix $\begin{bmatrix} 1 & 1 \\ 1 & 1 \end{bmatrix}$ invertible?

SOLUTION: This is the same argument that we hid in the footnote. If we multiply $\begin{bmatrix} 1 & 1 \\ 1 & 1 \end{bmatrix}$ by another matrix $\begin{bmatrix} a & b \\ c & d \end{bmatrix}$, we get:

$$\begin{bmatrix} 1 & 1 \\ 1 & 1 \end{bmatrix} \begin{bmatrix} a & b \\ c & d \end{bmatrix} = \begin{bmatrix} a+c & b+d \\ a+c & b+d \end{bmatrix}$$

and there is no solution to the equation $\begin{bmatrix} a+c & b+d \\ a+c & b+d \end{bmatrix} = \begin{bmatrix} 1 & 0 \\ 0 & 1 \end{bmatrix}$.

Here is a proof that there are six 2-by-2 invertible $\mathbf{F}_2$-matrices which does not require laborious listing. The first row cannot be $(0,0)$, so it has to be $(1,0)$, $(0,1)$ or $(1,1)$. Once we have the first row, the second row cannot be $(0,0)$ and cannot be identical to the first row, so we have two choices left for it. So there are a total of three choices for the first row times two choices each for the second row, leading to at most six possible invertible matrices. Here they are:

$$\begin{bmatrix} 1 & 0 \\ 0 & 1 \end{bmatrix}, \begin{bmatrix} 1 & 1 \\ 0 & 1 \end{bmatrix}, \begin{bmatrix} 1 & 0 \\ 1 & 1 \end{bmatrix}, \begin{bmatrix} 0 & 1 \\ 1 & 1 \end{bmatrix}, \begin{bmatrix} 0 & 1 \\ 1 & 0 \end{bmatrix}, \begin{bmatrix} 1 & 1 \\ 1 & 0 \end{bmatrix}. \tag{11.2}$$

Then you can check that each of these is really invertible.

If you know that A is invertible, there is a fairly straightforward algorithm for finding its inverse matrix. For sizes larger than 2-by-2, it is also too far off track to go into here. But here is the 2-by-2 theory, where the matrix entries are elements of a field. If

$$A = \begin{bmatrix} a & b \\ c & d \end{bmatrix}$$

we define the determinant of A, $\det(A)$ to be the number $ad - bc$. Then if $\det(A) \neq 0$,

$$A^{-1} = \begin{bmatrix} \dfrac{d}{ad - bc} & \dfrac{-b}{ad - bc} \\ \dfrac{-c}{ad - bc} & \dfrac{a}{ad - bc} \end{bmatrix}$$

which you can check by matrix multiplication: Both AA^{-1} and $A^{-1}A$ should equal the identity matrix $I = \begin{bmatrix} 1 & 0 \\ 0 & 1 \end{bmatrix}$. If $\det(A) = 0$, then A has no inverse.

You can now check that the six matrices listed in (11.2) all have determinant 1, and you can figure out each one's matrix inverse.

The General Linear Group of Invertible Matrices

THEOREM 11.3: The set of invertible n-by-n R-matrices form a group.

PROOF: Well, we said that the associative law holds. We have the neutral identity matrix I. And, by definition, every element in this set is invertible, so we have inverses. Are we done?

No. Now that we have limited ourselves to invertible matrices, we have to check point (1) on page 125 again. Is this new and smaller set of square invertible matrices still closed under matrix multiplication? The answer is yes.

If you desire a proof of that, here it is: Suppose A and B are invertible square matrices. Then A^{-1} and B^{-1} exist. We claim that the inverse of AB is $B^{-1}A^{-1}$. To check this, we must use the associative law several times: $(AB)(B^{-1}A^{-1}) = ((AB)B^{-1})A^{-1}) = (A(BB^{-1}))A^{-1} = (AI)A^{-1} = AA^{-1} = I$. Similarly, you can prove as an exercise that $(B^{-1}A^{-1})(AB) = I$. □

We need a symbol for this group: $\mathrm{GL}(n, R)$. The "GL" stands for the words "general linear," which are a historical artifact of the theory.[5]

[5]The adjective "general" meant that the determinant is allowed to be general, that is, not necessarily equal to 1.

DEFINITION: $\mathrm{GL}(n, R)$ is the group of all n-by-n invertible R-matrices, where the group law is given by matrix multiplication.

For example, the six matrices (11.2) form the group $\mathrm{GL}(2, \mathbf{F}_2)$.

Another thing to remember about matrix multiplication is that it usually is not commutative. For example, in $\mathrm{GL}(2, \mathbf{Q})$ we have

$$\begin{bmatrix} 1 & 2 \\ 3 & 4 \end{bmatrix}\begin{bmatrix} 5 & 6 \\ 7 & 8 \end{bmatrix} = \begin{bmatrix} 19 & 22 \\ 43 & 50 \end{bmatrix} \quad \text{and} \quad \begin{bmatrix} 5 & 6 \\ 7 & 8 \end{bmatrix}\begin{bmatrix} 1 & 2 \\ 3 & 4 \end{bmatrix} = \begin{bmatrix} 23 & 34 \\ 31 & 46 \end{bmatrix}.$$

Another example drawn from $\mathrm{GL}(2, \mathbf{F}_2)$:

$$\begin{bmatrix} 1 & 1 \\ 0 & 1 \end{bmatrix}\begin{bmatrix} 0 & 1 \\ 1 & 0 \end{bmatrix} = \begin{bmatrix} 1 & 1 \\ 1 & 0 \end{bmatrix} \quad \text{and} \quad \begin{bmatrix} 0 & 1 \\ 1 & 0 \end{bmatrix}\begin{bmatrix} 1 & 1 \\ 0 & 1 \end{bmatrix} = \begin{bmatrix} 0 & 1 \\ 1 & 1 \end{bmatrix}.$$

The Group GL(2, Z)

The group $\mathrm{GL}(2, \mathbf{Z})$ is one way in which number theory and group theory can come together, though we can give only a few details in this book. For example, the theory of modular forms, referred to in chapters 21–23, depends crucially on $\mathrm{GL}(2, \mathbf{Z})$ and its properties. In this section, we will see that even the simple problem of enumerating the elements of $\mathrm{GL}(2, \mathbf{Z})$ depends on elementary number theory.

We know that $\mathrm{GL}(2, \mathbf{Z})$ is the set of 2-by-2 matrices with integer entries and determinant equal to 1 or -1. (This is because 1 and -1 are the only elements in $\mathbf{Z}$ that have multiplicative inverses in $\mathbf{Z}$.) It is not that hard to find elements of $\mathrm{GL}(2, \mathbf{Z})$, after you know that the determinant of $A = \begin{bmatrix} a & b \\ c & d \end{bmatrix}$ is $ad - bc$. The entries a, b, c, and d must all be integers, and the only requirement for invertibility is that $ad - bc = 1$ or $ad - bc = -1$. We investigate the case where $ad - bc = 1$. Then A^{-1} is given by $\begin{bmatrix} d & -b \\ -c & a \end{bmatrix}$. For example, we have

$$\begin{bmatrix} 3 & 5 \\ 4 & 7 \end{bmatrix}^{-1} = \begin{bmatrix} 7 & -5 \\ -4 & 3 \end{bmatrix}.$$

How do we construct matrices in $GL(2, \mathbf{Z})$? We start by picking the first row at random: say $[6 \quad 15]$. Then we look for the second row $[c \quad d]$ to satisfy $6d - 15c = 1$. Whoops. That can never happen if c and d are integers, because the left-hand side is an exact multiple of 3, and so it could never equal 1. We had better make sure that a and b share no common factor. Two numbers like that are called *relatively prime*. For example, 3 and 20 are relatively prime, 100 and 57 are relatively prime, but 37 and 111 are not relatively prime, because both are multiples of 37. (Yes, a number is always considered to be a multiple of itself.)

Armed with this knowledge, we now pick our first row so the two entries in it are relatively prime: say $[6 \quad 11]$. Now we look for $[c \quad d]$ to satisfy $6d - 11c = 1$. One way to do this is to list a lot of multiples of 6 and 11 and look for entries in the list that are only one apart:

$$6, 12, 18, 24, 30, 36, 42, 48, 54, 60, 66, 72, 78, 84, 90, \ldots$$

$$11, 22, 33, 44, 55, 66, 77, 88, 99, 110, \ldots$$

We immediately notice the 11 on the bottom row and the 12 on the top row, which leads us to choose $c = 1$ and $d = 2$. Check: $6 \times 2 - 11 \times 1 = 1$. We also see 78 on the top list and 77 on the bottom, leading to the choice of $c = 7$ and $d = 13$. Check: $6 \times 13 - 11 \times 7 = 78 - 77 = 1$. If we are clever, we can even exploit the 55 on the bottom row and 54 on the top row. The problem is that $54 - 55 = -1$, not 1. But we can solve this by multiplying through by -1: in fact $-54 - (-55) = 1$. This leads to the choice $c = -5$ and $d = -9$. Check: $6 \times (-9) - 11 \times (-5) = -54 - (-55) = 1$. We thus get the following matrices in $GL(2, \mathbf{Z})$:

$$\begin{bmatrix} 6 & 11 \\ 1 & 2 \end{bmatrix}, \quad \begin{bmatrix} 6 & 11 \\ 7 & 13 \end{bmatrix}, \quad \begin{bmatrix} 6 & 11 \\ -5 & -9 \end{bmatrix}.$$

We see that not only have we succeeded, but we have also found several solutions to our problem. This illustrates a general theorem: If a and b are relatively prime integers, then there are infinitely many pairs of integers c and d with the property $ad - bc = 1$.

Therefore, there are also infinitely many matrices of determinant 1 in $\mathrm{GL}(2, \mathbf{Z})$ whose top row is $[a \quad b]$.[6]

Solving Matrix Equations

We now go back and look at solving systems of linear equations with rational coefficients. It is fun to view this problem from the point of view of $\mathrm{GL}(n, \mathbf{Q})$. Suppose we have a system of n **Z**-equations with n unknowns, and no exponent greater than 1: a "linear system." As we explained at the end of chapter 10, we can write this system using matrices. If we let X be the n-by-1 matrix of unknowns, we can find an n-by-n **Q**-matrix A and an n-by-1 **Q**-matrix B so that our system is expressed by the single matrix equation[7]

$$AX = B. \tag{11.4}$$

Everything now depends on whether or not A is in $\mathrm{GL}(n, \mathbf{Q})$. If it is, then A has a matrix inverse A^{-1}. If we multiply both sides of (11.4) on the left by A^{-1}—be careful to multiply *on the left*, because matrix multiplication is not commutative—we get

$$A^{-1}(AX) = A^{-1}B.$$

Matrix multiplication is associative, so this is equivalent to

$$(A^{-1}A)X = A^{-1}B.$$

And $A^{-1}A = I$ and $IX = X$ so this is equivalent to

$$X = A^{-1}B. \tag{11.5}$$

Do you see what this says? First of all, there is a solution, and equation (11.5) is telling us what it is: $X = A^{-1}B$. Also it is telling us that the solution is unique. There is no choice. Equation (11.4) is

---

[6]There is the Euclidean Algorithm, which gives you all the pairs of c's and d's to go with a given pair (a, b). You can find this in any number theory textbook, but we do not need it for our purposes.

[7]A and B are actually **Z**-matrices, but since **Z** is contained in **Q**, we may consider them as **Q**-matrices. This allows us to look for fractional solutions.

equivalent to equation (11.5), so if you want to solve equation (11.4) you have to take $X = A^{-1}B$.

Remember that this was all predicated on the assumption that A was invertible, that is, A in $\mathrm{GL}(n, \mathbf{Q})$. If A is not invertible, then either (11.4) has no solution at all, or else it has infinitely many solutions.

EXAMPLE: We solve the simultaneous set of equations

$$2x + 3y + 4z = 12. \tag{11.6}$$

$$x + 0y + 2z = 6. \tag{11.7}$$

$$3x + 4y + 8z = 0. \tag{11.8}$$

Go back to the exercise on page 127 involving the matrices A and H. If we let X be the matrix

$$\begin{bmatrix} x \\ y \\ z \end{bmatrix}$$

and B be the matrix

$$\begin{bmatrix} 12 \\ 6 \\ 0 \end{bmatrix}$$

then the system (11.6)–(11.8) can be written as one equation

$$AX = B.$$

In the exercise above, we proved that $H = A^{-1}$. So from (11.5) we see that our system has a unique solution, namely $X = HB$. If you perform the actual matrix multiplication HB, you will find that

$$X = \begin{bmatrix} 24 \\ 0 \\ -9 \end{bmatrix}.$$

In other words, the unique solution is $x = 24$, $y = 0$, $z = -9$.

By the way, it is just a coincidence that there are no fractions in the solution, even though there are fractions in H. This is because all the integers on the right-hand side of our system are divisible by the denominators in H. To convince yourself of this, try to solve the system

$$2x + 3y + 4z = 12.$$

$$x + 0y + 2z = 6.$$

$$3x + 4y + 8z = 1.$$

SOLUTION: $x = 77/3, y = -1/3, z = -28/3$.

The advantage of knowing H is that it is now easy to solve any system such as (11.6)–(11.8), where the left-hand sides are the same but the right-hand side B' might be different. You just matrix-multiply HB'. The experts will tell you, however, that if you just want to solve one system, not a whole bunch with the same left-hand sides, then the fastest thing to do is use something called "Gaussian elimination." (You can read about it in those linear algebra textbooks.)

Chapter 12

GROUP REPRESENTATIONS

Road Map

Before diving back into number theory, we explain the key concept that puts together groups, permutations, and matrices: *group representations*. We will give several different examples. Except for the last example, which uses elliptic curves, we mostly keep away from number theory in this chapter, so the unadorned concept of group representation can stand out more clearly.

Our central example is that of the symmetries of a tetrahedron, which shows how geometrical symmetry can be interpreted using group theory, and then be understood further using representation theory. This example is a simple version of many other, more complicated, geometrical symmetry groups and their representations, which could be the subject matter of a different book.

Morphisms of Groups

At last, we come to the heart of this book, or at least the pericardium. A *group representation* is nothing more nor less than a morphism from one group to another group. The reason we use the special term "representation" is that the target group is chosen to be one we are especially comfortable with, or one whose properties are important for understanding the source group. The two types of representations we will look at are *permutation representations* and *linear representations*.

What is a morphism from the group H to the group K? It is a function, call it $f(x)$, for example, from H to K with just one property:

$$f(x \circ y) = f(x) \circ f(y) \tag{12.1}$$

for any two elements x and y in H. In other words: f is a rule that assigns to any element, say x, from H, exactly one element $f(x)$ from K. And it does this in such a way that (12.1) holds. Note that in (12.1), the little circle in $x \circ y$ refers to the group law in H, while that in $f(x) \circ f(y)$ refers to the group law in K. We summarize all of this as follows: $f : H \to K$ is a morphism of groups.

EXAMPLE: We start with an example that captures three of the most basic facts about multiplication of numbers:

- positive × positive = positive.
- positive × negative = negative.
- negative × negative = positive.

In this example, our source group H is $\mathbf{R}^{\times}$, the set that contains all real numbers other than 0, with the group law of multiplication. The target group K is $\{+1, -1\}$, again with the group law of multiplication. The morphism f is the following rule:

$$f(x) = \begin{cases} +1 & \text{if } x > 0. \\ -1 & \text{if } x < 0. \end{cases}$$

You can check that the equation $f(xy) = f(x)f(y)$ will always be true, if x and y are any nonzero real numbers.

In this equation, we throw away a lot of information about the source group $\mathbf{R}^{\times}$ and preserve only the sign of the number. Our representation aggregates the source group into two clumps—positive numbers and negative numbers—and ignores all of the other information about real numbers. It captures the rules for multiplication of signed numbers at the expense of forgetting about all of the other properties of real numbers.

This observation leads us to another definition. The morphism f in this example ignored a lot of information about the group $\mathbf{R}^{\times}$ in order to represent a single bit of information. The technical terminology is that this particular representation was not *faithful*. To define "faithful," first notice that if $r : H \to K$ is any morphism of groups, and e is the neutral element of H, then $r(e)$ must be the neutral element of K. This is because of two facts:

1. The only element of a group satisfying the equation $X \circ X = X$ is the neutral element.
2. Because r is a morphism, $r(e) \circ r(e) = r(e \circ e) = r(e)$.

DEFINITION: If H and K are groups with neutral elements e_H and e_K, and r a morphism of H to K, then we say that r is *faithful* if the only solution to the equation $r(x) = e_K$ is given by $x = e_H$.

This definition, along with a little algebra, implies something more comprehensive: If r is a faithful representation, and x and y are two different elements of the group H, then the elements $r(x)$ and $r(y)$ in K will not be equal.[1] Incidentally, you can see from the definition that our first example is not faithful, because there are many real numbers x that solve the equation $f(x) = 1$ (namely, all of the positive real numbers).

EXAMPLE: Let $H = \mathbf{Z}$ with the group operation of addition (so that $3 \circ 8 = 11$). Let $K = \mathbf{F}_2$ with the group law of addition (so that $1 \circ 1 = 0$). Let $f(x)$ be the rule:

$$f(x) = \begin{cases} 1 & \text{if } x \text{ is odd.} \\ 0 & \text{if } x \text{ is even.} \end{cases}$$

[1]PROOF: We prove the contrapositive. Suppose that g and h are two unequal elements of H, but $r(g) = r(h)$. We will show that the equation $r(x) = e_K$ has at least one solution other than e_H. By the axioms of a group, we know that because $g \neq h$, there is an element k in H not equal to e_H with the property that $h = g \circ k$. Then by (12.1), $r(h) = r(g) \circ r(k) = r(h) \circ r(k)$ which implies, by cancellation of $r(h)$ on the left, that $r(k) = e_K$.

You can then check that (12.1) is true. So f is a *morphism* from $\mathbf{Z}$ to $\mathbf{F}_2$. This representation is also not faithful, because there are lots of solutions to the equation $f(x) = 0$.

This group representation captures the following facts about addition:

- even + even = even.
- even + odd = odd.
- odd + odd = even.

EXAMPLE: Let H be the group of positive real numbers, with the group law multiplication. Let K be the group of all real numbers, with the group law addition. Let $f(x) = \log_{10} x$. Then f is a morphism, because $\log_{10}(xy) = \log_{10}(x) + \log_{10}(y)$.

This representation *is* faithful. One way to see that is that there is only one solution to the equation $\log_{10} x = 0$, namely, $x = 1$. Another way is to use the other property we mentioned, and notice that if $\log_{10} x = \log_{10} y$, then $x = y$.

The examples of morphisms in this book, from now on, will have other kinds of groups, rather than number systems, as their source or target. Many of our examples will not be faithful representations. At first sight, it might be surprising that a representation that is not faithful is a useful one. In fact, a representation that is not faithful emphasizes certain features of the source group and obscures others, enabling us to understand better the source group. After this chapter, we will mostly use the absolute Galois group G as our source group. Even though G is infinite, typically our target groups will be finite (and hence simpler to understand). Any morphism from an infinite group to a finite group cannot be faithful, and hence will focus on certain features of G at the expense of others. By using many representations, even though none of them will be faithful, we can hope gradually to understand more and more of the facets of G.

DEFINITION: A *permutation representation* of a group H is a morphism from H to a permutation group Σ_A, where A is some set.

DEFINITION: A *linear representation* (or *matrix representation*) of a group H is a morphism from H to a matrix group $\mathrm{GL}(n, R)$, where R is some number system.

For instance, suppose that G is the absolute Galois group of $\mathbf{Q}$. Look back at our discussion in chapter 8 about how any element g in G permutes the roots of any $\mathbf{Z}$-polynomial. So if we pick a particular polynomial, we can ask how any element of G permutes its roots. We will explain this example in detail in chapter 14. You can also look at our description of the n-torsion of an elliptic curve in chapter 9. There, too, we can think of an element of G as permuting the n-torsion. In this case, we can also produce linear representations of G. See the end of this chapter and chapter 18 for more details.

A_4, Symmetries of a Tetrahedron

This is a very long example. If the going gets a bit dense, don't worry; we will not need the details of this example in any other part of this book. Our source group in this example is a permutation group called A_4. First, we will tell you about the group abstractly, and then we will give you some other ways to think about it by using representation theory.

Our group A_4 is a group of permutations of the set $\{1, 2, 3, 4\}$, but it does not contain every permutation (or else it would be $\Sigma_{\{1,2,3,4\}}$). In fact, A_4 contains exactly 12 permutations, or half of all of the permutations in $\Sigma_{\{1,2,3,4\}}$.

To start, A_4 contains every permutation that leaves one number fixed, and cyclically permutes the other three numbers. For example, A_4 contains the permutation that sends 2 to 2, and cyclically permutes 1, 3, and 4, meaning that $1 \to 3 \to 4 \to 1$:

$$1 \to 3$$

$$2 \to 2$$

$$3 \to 4$$

$$4 \to 1.$$

The short notation for writing this permutation is (134), where the omission of the number 2 means that 2 is not changed by the permutation.

There are eight such permutations.

EXERCISE: List all eight of the permutations that fix one element of $\{1, 2, 3, 4\}$ and cyclically permute the other three numbers.

SOLUTION: It would consume too much space to list these permutations using the arrow notation. Using the shorter cycle notation, the eight permutations are (123), (132), (134), (143), (124), (142), (234), and (243).

Remember that a group must also contain the neutral element e (which leaves every element of $\{1, 2, 3, 4\}$ fixed), so you now know nine elements of A_4. The other three elements can be found because our group must be closed under the composition law.

The result of $(123) \circ (124)$ (which, you remember, means first doing the permutation (124) and then the permutation (123)) is:

$$1 \to 3$$

$$2 \to 4$$

$$3 \to 1$$

$$4 \to 2.$$

The way to write this in cycle notation is (13)(24).

There are two other permutations that result from combining the elements of our group: (12)(34) and (14)(23). Here is the list of all of the permutations in A_4: e, (123), (132), (134), (143), (124), (142), (234), (243), (12)(34), (13)(24), (14)(23).

EXERCISE: Find a permutation of $\{1, 2, 3, 4\}$ that is not in A_4.

SOLUTION: Here are all of the permutations in $\Sigma_{\{1,2,3,4\}}$ that are not in A_4: (12), (13), (14), (23), (24), (34), (1234), (1324), (1423), (1243), (1342), and (1432).

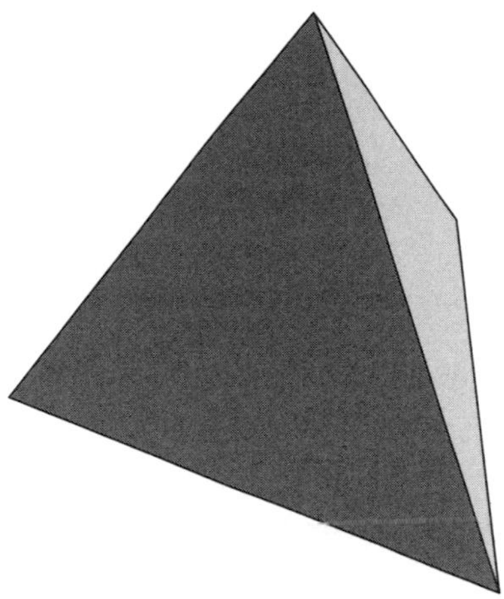

Figure 12.1: A regular tetrahedron

This is pretty abstract stuff. Fortunately, there is another way to think about A_4.

DEFINITION: A *regular tetrahedron* is a solid figure with four congruent equilateral triangles as its faces.

You can see a picture of a tetrahedron in Figure 12.1.

Number the vertices of the tetrahedron from 1 to 4. For the sake of illustration, label the top vertex with the number 1, and the three on the base 2, 3, and 4 (counterclockwise).

Now imagine rotating the tetrahedron so that when you are done, the tetrahedron still looks the same, except that the vertices may have moved around. For example, you may keep the vertex numbered 1 fixed, and spin the tetrahedron 120° counterclockwise, which will shift the other three vertices. The effect of this spinning on the vertices is the permutation

$$1 \to 1$$

$$2 \to 3$$

$$3 \to 4$$

$$4 \to 2.$$

In other words, this spin of the tetrahedron can be thought of as producing an element of A_4.

In fact, every element of A_4 comes from some way of spinning the tetrahedron. You may have some difficulty finding the rotation that corresponds to the permutation (12)(34), but it does exist. You can find it if you play long enough with a physical model of a tetrahedron.

What we have done is to represent A_4 as permutations of the vertices of a tetrahedron. Each element of A_4 gives a different permutation. There is no way to permute the vertices of the tetrahedron via a rotation other than by the permutations listed in A_4. In other words, the permutation (12), which flips vertices 1 and 2 and leaves vertices 3 and 4 unchanged, is not a possible permutation of the vertices that can be achieved by rotating the tetrahedron, nor are any of the other 11 permutations listed above as elements of $\Sigma_{\{1,2,3,4\}}$ that are not in A_4.

We have thus given our abstract group A_4 a more physical interpretation. The classical terminology is that A_4 is the group of *orientation-preserving symmetries* of the tetrahedron. Now we can also make a connection with ideas mentioned earlier. Look back at the definition of the group SO(3) in chapter 2. It is the group of rotations of a rigid sphere.

Representations of A_4

We now take our ivory sphere out of its spherical box, erasing whatever dots and lines we have drawn on it so it is blank, and carefully draw on it a perfect tetrahedron. This means putting four dots on the sphere that are all equally spaced from one another. See Figure 12.2.

Put the sphere back into its container, and on the inside of the container put four dots touching the dots on the sphere, and number them the same way, 1, 2, 3, 4, so that dot 1 touches dot 1, dot 2 touches dot 2, and so on. Now you are ready to play.

Rotate the sphere inside the box. You are applying a group element from SO(3). Unless you are very careful, the four dots

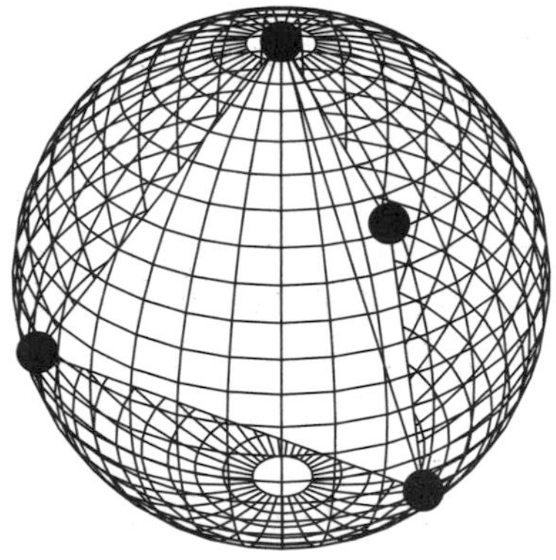

Figure 12.2: A tetrahedron with a sphere

on the sphere will no longer exactly touch the four dots on the box. So be careful when doing this. We collect together in a set all the elements of SO(3) that do end up with the dots on the sphere touching the dots on the box, possibly in a new way. For example, if you rotate the sphere around the axis that goes through dot 1 and the center of the sphere exactly 120°, the dots will line up, as we noticed above.

Take an orange or a ball and try this out. How many different rotations can you find that keep the dots lined up? Remember that one possibility is the neutral rotation e which leaves everything in place.

You should be able to convince yourself that each element of A_4 tells how to spin the tetrahedron around, and each of those rotations gives us an element of SO(3). What we have found is a way to think of elements of A_4 as determined by rotations in SO(3).

The important thing is that this is a morphism. This morphism is a function that we call r. If σ is a permutation in A_4, then $r(\sigma)$ is the rotation of the sphere that comes from spinning the tetrahedron to permute the vertices the way that σ tells us to. So r is a function from A_4 to SO(3). The morphism property means that if τ is another permutation in A_4, then $r(\sigma \circ \tau) = r(\sigma) \circ r(\tau)$. In words: If you first do τ to $\{1, 2, 3, 4\}$, then do σ, and then see what rotation performs that composite permutation of the numbered dots, that is the same as the rotation you get by first doing the rotation $r(\tau)$ and then the rotation $r(\sigma)$.

This is almost an example of a linear representation, but we are not there yet. Remember that we defined a linear representation as one whose target is a matrix group. So we will tell you how to find a representation of all of SO(3) whose target is the matrix group $\mathrm{GL}(3, \mathbf{R})$, the group of invertible 3-by-3 real matrices. Remember that every element of SO(3) is a rotation. Now we can associate a 3-by-3 matrix to any rotation g.

Take our beautiful sphere of radius 1 and center it at the origin of a three-dimensional coordinate axis, where the z-axis is pointing up. On the x-axis, mark off the point $P = (1, 0, 0)$. This is the point on the x-axis one unit from the origin. On the y-axis, mark off the point $Q = (0, 1, 0)$. This is the point on the y-axis one unit from the origin. On the z-axis, mark off the point $R = (0, 0, 1)$. This is the point on the z-axis one unit from the origin. Fix the sphere's starting position so that dot number 1 coincides with R on the z-axis.[2]

Now apply the rotation g to the sphere *and to the coordinate system that is stuck in it*. The whole set of axes rotates. But we also imagine that a copy of the original set of axes stays in place, for measurement purposes. When we do this, P, Q, and R will rotate to three new points P', Q', and R'. Each of these new points has coordinates, say $P' = (a, b, c)$, $Q' = (d, e, f)$, and $R' = (h, k, j)$.

For example, if g is rotation by 120° counterclockwise (looking down) around the axis that goes through dot number 1 and the origin, then $P' = (-1/2, \sqrt{3}/2, 0)$, $Q' = (-\sqrt{3}/2, -1/2, 0)$, and $R' = (0, 0, 1)$. (Because R stays fixed under g, $R' = R$.)

We associate to the rotation g the matrix

$$s(g) = \begin{bmatrix} -\frac{1}{2} & -\frac{\sqrt{3}}{2} & 0 \\ \frac{\sqrt{3}}{2} & -\frac{1}{2} & 0 \\ 0 & 0 & 1 \end{bmatrix}.$$

[2]The point P will coincide with the dot numbered 1, but the other three dots will lie *below* the xy-plane. The points P, Q, and R all coincide with the surface of the sphere, because the sphere has a radius of one unit.

We get the three columns of this matrix by writing vertically the entries in P', Q', and R'. The same prescription serves to define $s(h)$ for any rotation h. So if we let h run through the rotations that permute the vertices of the tetrahedron, s defines a function from A_4 to $\mathrm{GL}(3, \mathbf{R})$. In fact, this function is a morphism, and so it is a linear representation of A_4. In this linear representation, the permutation (234) is represented by the matrix

$$\begin{bmatrix} -\frac{1}{2} & -\frac{\sqrt{3}}{2} & 0 \\ \frac{\sqrt{3}}{2} & -\frac{1}{2} & 0 \\ 0 & 0 & 1 \end{bmatrix}.$$

Two remarks:

1. This same construction lets us define s on *any* rotation in SO(3). This gives us a linear representation from SO(3) to $\mathrm{GL}(3, \mathbf{R})$. So you see that the groups that get represented do not have to be finite.
2. SO(3) is a noncommutative group. In fact, even A_4 is noncommutative. This is one reason why it is important to have noncommutative targets available for representations. Otherwise, we would never be able to capture noncommutative information about a source group.

But we do not have to work with rotations to come up with an interesting linear representation of A_4. In fact, we can get a four-dimensional linear representation of all of $\Sigma_{\{1,2,3,4\}}$ in the following way.

For any permutation π of $\{1, 2, 3, 4\}$, set $V(\pi)$ to be the matrix you get by starting with the 4-by-4 zero matrix[3] and then replacing exactly four of the zeros by ones. Which ones? Put a 1 in the first column and the $\pi(1)$th row; put a 1 in the second column and the $\pi(2)$th row; put a 1 in the third column and the $\pi(3)$th row; and, finally, put a 1 in the fourth column and the $\pi(4)$th row.[4]

[3]A *zero matrix* is one whose entries are all 0.

[4]Remember that a permutation is a function, so we can use the functional notation $\pi(x)$ for the value of the permutation π on x, that is, the result of the permutation π acting on x.

For example, suppose that we start with the permutation

$$1 \to 1$$
$$2 \to 3$$
$$3 \to 4$$
$$4 \to 2.$$

This corresponds to the matrix

$$V(\pi) = \begin{bmatrix} 1 & 0 & 0 & 0 \\ 0 & 0 & 0 & 1 \\ 0 & 1 & 0 & 0 \\ 0 & 0 & 1 & 0 \end{bmatrix}$$

in GL(4, $\mathbf{R}$). For example, because $\pi(4) = 2$, there is a 1 in the fourth column and the second row.

We have now defined a morphism V from $\Sigma_{\{1,2,3,4\}}$ to GL(4, $\mathbf{R}$). (It is tedious to check that this is a morphism, but it is not difficult.) But you could view these 0's and 1's as being elements of any number system. So V can be viewed as a linear representation into GL(4, $\mathbf{Z}$) or GL(4, $\mathbf{F}_p$) for any prime p. By the way, this representation is obviously faithful.

Mod *p* Linear Representations of the Absolute Galois Group from Elliptic Curves

We call a linear representation of a group to GL(n, $\mathbf{F}_p$) a "mod p" representation for short. Here is an example of an interesting "mod p" representation. It is a representation of the absolute Galois group G, and we can get it from the p-torsion of elliptic curves, but first we need to know a bit more about the torsion.

THEOREM 12.2: Pick an elliptic curve E and a positive integer n. Let $E[n]$ be the set of all n-torsion.[5] Then we can choose two particular elements P and Q inside of $E[n]$ so that

[5]The n-torsion $E[n]$ was defined on page 112. Remember that $E[n]$ has n^2 elements.

every element in $E[n]$ can be written as $aP + bQ$, where a and b run separately over integers from 0 to $n - 1$.

Remember that aP means $\overbrace{P + P + \cdots + P}^{a \text{ times}}$ if a is not 0, whereas it means $\mathcal{O}$, the neutral element of the elliptic curve, if $a = 0$. Also, remember that we have already mentioned on page 113 how to use the n-torsion to get a permutation representation: We pick any element g of G, and we think about how g permutes the n^2 elements of $E[n]$.

Now we look at $g(P)$ and $g(Q)$. We know that $g(P)$ and $g(Q)$ must be elements of $E[n]$, so we can write $g(P) = aP + bQ$ and $g(Q) = cP + dQ$, where a, b, c, and d are all numbers between 0 and $n - 1$. Then our representation applied to g is defined to be the matrix

$$r(g) = \begin{bmatrix} a & c \\ b & d \end{bmatrix}.$$

Now suppose $n = p$ is a prime number and view the matrix $r(g) = \left[\begin{smallmatrix} a & c \\ b & d \end{smallmatrix}\right]$ as an element of $\mathrm{GL}(2, \mathbf{F}_p)$. You can check that if g_1 and g_2 are two elements of G, then $r(g_1 \circ g_2) = r(g_1)r(g_2)$. You have to use the fact that $g(\ell P + mQ) = \ell g(P) + mg(Q)$ for any integers ℓ and m.[6] In any case, r is an honest-to-goodness linear representation.

This representation r has been described in a very abstract way. We have told you how in theory you can find the matrix $r(g)$. In actual practice, it can be hard to do. One of the beauties of the modern theory of elliptic curves is that a tremendous amount of information can be proved about the representation r, without having to write it down explicitly in formulas. Instead, mathematicians exploit all the symmetries and relationships implicit in the definitions of elliptic curves and Galois groups. We will give an example of one type of information that is known about r, after we define Frobenius elements in G in chapter 16.

[6]This is true because all the algebra used to find $\ell P + mQ$ involves only rational numbers, and so is unchanged when you apply the Galois element g.

Remark for lovers of trivia: If H is any group, n any positive integer, and R any number system, there is always the "trivial" representation from H to $\mathrm{GL}(n, R)$. It sends every element of H to the identity matrix I. This representation is as far from faithful as it could be, because every single element in H gets sent to the same element in $\mathrm{GL}(n, R)$.

What good is the trivial representation? It may come up in a theorem where some representation (very abstractly defined) is proved to be equal to the trivial representation. Or we may want to say some representation is nontrivial. So we need to be able to talk about the trivial representation. You may as well ask, "What good is 0?"

For example, consider the representation $r : G \to \mathrm{GL}(2, \mathbf{F}_p)$ defined as above from the p-torsion of an elliptic curve E. It can be proved that r is trivial if and only if *all* of the p-torsion elements of $E(\mathbf{C})$ actually lie in $E(\mathbf{Q})$, that is, have *rational* coordinates. This clearly would be an important fact about the p-torsion of E.

In all cases, mod p representations of the absolute Galois group G are not faithful. How can they be? G is infinite, but $\mathrm{GL}(2, \mathbf{F}_p)$ is finite. Many different elements of G have to go to the same matrix. This is no big deal. It just means we will not be able to get complete information about G from any single mod p representation.

Chapter 13

THE GALOIS GROUP OF A POLYNOMIAL

Road Map

The absolute Galois group G is very large and difficult to understand. We can glimpse various aspects of it by constructing permutation representations of G that have as a target some finite permutation group. It is like trying to understand an infinite universe by looking at photos of small pieces of it—a nebula here, a supernova there. Figuring out how these photos piece together is very difficult. The Reciprocity Laws to be discussed in chapter 17 are one of the ways of obtaining a bigger picture.

In this chapter we discuss the individual "photos" themselves. Each one is a Galois group of a **Z**-polynomial. In the next chapter we will glue these onto G and hence to each other via "the restriction morphisms." We will see that the infinite set of all possible "photos" does cover the whole of G.

The Field Generated by a Z-Polynomial

Recall that $\mathbf{Q}^{\text{alg}}$ is the set of all roots of all **Z**-polynomials. We have called the group of all permutations of $\mathbf{Q}^{\text{alg}}$ that preserve addition, subtraction, multiplication, and division the "absolute Galois group of **Q**." (See chapter 8.) We have agreed to designate it by the letter G. It is a very large—in fact, infinite—group. In order to get a grasp on it, we are going to discuss a lot of smaller groups, also called Galois groups.

Each of these groups is the Galois group of a **Z**-polynomial. Here is the idea. Pick a **Z**-polynomial $f(x)$. Instead of looking at all of $\mathbf{Q}^{\text{alg}}$, we only look at the part we get by taking all of the roots of $f(x)$ (all contained in $\mathbf{Q}^{\text{alg}}$) and all of their arithmetic combinations. If $f(x)$ has degree d, then it will have at most d roots. By making arithmetic combinations of them, we get a whole bunch of numbers called a "field." Follow this recipe:

Start with a big pot. Take a particular **Z**-polynomial $f(x)$ of degree $d \geq 1$. Then throw into the pot all of the roots of $f(x)$, along with all rational numbers. Then stir. Stirring means we form all possible sums, differences, products, and quotients of all the numbers in the pot. Then stir again, and keep doing this over and over again, watching the brew grow—double, double, toil, and trouble. After stirring infinitely many times, we call the result $\mathbf{Q}(f)$—the field generated by the roots of $f(x)$.[1]

EXAMPLE: If $f(x)$ is the polynomial $(x^2+1)(x^3-7)$, some of the numbers you get in the pot after enough stirring will be $\frac{3}{4}$, i, $\sqrt[3]{7}$, $\frac{i+\sqrt[3]{7}-\frac{3}{4}}{77}$, $\frac{i\sqrt[3]{7}+i^5\sqrt[3]{7}^2}{3i-4\sqrt[3]{7}+8}$, and so on.

There are other ways to describe $\mathbf{Q}(f)$. First of all, it is a *field*. Recall the definition of a field. It is a set of numbers *closed* under addition, subtraction, multiplication, and division. This means that if u and v are any two elements of the field, then $u+v$, $u-v$, uv and $\frac{u}{v}$ are also in the field (in the case of $\frac{u}{v}$, we assume that $v \neq 0$). Secondly, $\mathbf{Q}(f)$ is the *smallest* field that contains all the roots of $f(x)$ and all rational numbers.[2]

Now we have the field $\mathbf{Q}(f)$, which is a subset of $\mathbf{Q}^{\text{alg}}$. We can look at the permutations of $\mathbf{Q}(f)$ that preserve all the arithmetic operations. This set of permutations is a group, called the Galois group of $f(x)$, denoted $G(f)$. The group $G(f)$ is always finite, which

---

[1]A theorem tells us that the contents actually will stop growing after some finite number of stirrings.

[2]All rational numbers are in $\mathbf{Q}(f)$ because by definition we threw them into the pot at the start. However, it is interesting to note that if $f(x)$ has at least one nonzero root, say u, then we can build up all of the rationals from u alone by forming $\frac{u}{u}=1$, then $1+1+\cdots+1$ to get all of the positive integers, then $1-1$ to get 0, $0-1-1-\cdots-1$ to get the rest of the integers, and then $\frac{a}{b}$ to get all of the rational numbers.

means it must be a lot simpler than the big Galois group G. In the next chapter, we will see some connections between all of these groups. (It may be a good idea to review chapter 3 at this time.)

Why is $G(f)$ finite? The reason is that any permutation in $G(f)$ must permute the roots of f, and there are at most d different roots. After you know how the roots are permuted, this fixes how all the other numbers in $\mathbf{Q}(f)$ must be permuted, because they are cooked, using the operations of basic arithmetic, only out of these roots and $\mathbf{Q}$—no other ingredients. And elements in $G(f)$, just like elements of G, must fix every rational number, as we saw on page 97. So there are at most $d!$ elements in $G(f)$.

We summarize with the following theorem:

THEOREM 13.1: Let f be a $\mathbf{Z}$-polynomial and let A be the set of all its roots in $\mathbf{Q}^{\text{alg}}$. Let r be the function from $G(f)$ to Σ_A which sends any element of $G(f)$ to the permutation it causes on A. Then r is a faithful morphism and hence a faithful permutation representation of $G(f)$.

In fact, it is useful at this point to redefine $G(f)$ as the set of permutations of the roots of f that, when extended to $\mathbf{Q}(f)$, preserve addition and multiplication. Thus:

DEFINITION: Let f be a $\mathbf{Z}$-polynomial and let A be the set of all its roots. Then:

- $\mathbf{Q}(f)$ is the smallest field containing $\mathbf{Q}$ and all of the elements of A.
- $G(f)$ is the set of permutations σ of A with the following property: There is some permutation of $\mathbf{Q}(f)$ that preserves addition and multiplication and that is the same as σ on the elements of A.

Examples

Phew! Maybe some examples will help.

EXAMPLE: $f(x) = x - 3$. This is a really easy example. At the beginning, only the root(s) of $f(x)$ and the rational numbers $\mathbf{Q}$

are in the pot. The only root of $f(x)$ is 3, so our pot contains only all of $\mathbf{Q}$. If we start adding, subtracting, multiplying, and dividing elements of $\mathbf{Q}$ (where we, of course, remember not to divide by 0), we only get other rational numbers. So $\mathbf{Q}(f)$ is just $\mathbf{Q}$. Any permutation of $\mathbf{Q}$ that preserves arithmetic must preserve 0 and 1 and hence all integers and hence all rational numbers. Such a permutation can only be the identity permutation. So $G(f)$ is the "trivial group" that only has one element in it, the neutral element. We generally call the neutral element in a group "e," but from now on, we will call the neutral element in a Galois group "ι," the Greek letter iota. It stands for "identity."

EXAMPLE: $f(x) = x^2 - 5$. This one is not too difficult. The two roots of $f(x)$ are $\sqrt{5}$ and $-\sqrt{5}$. If you multiply by various rational numbers and add to other rational numbers you find that $\mathbf{Q}(f)$ contains every number of the form $a + b\sqrt{5}$, where a and b are arbitrary rational numbers. Some more work will convince you that this is all there is in $\mathbf{Q}(f)$.

Now we try to figure out what $G(f)$ is. Of course, $G(f)$ always possesses the neutral element ι. If σ is another element of $G(f)$, what could it do to $\sqrt{5}$? It must take it to one of the roots of $f(x)$, so either to $\sqrt{5}$ or to $-\sqrt{5}$. Remember that any element of any Galois group fixes every rational number. If σ takes $\sqrt{5}$ to $\sqrt{5}$, then because it respects all arithmetic operations, it will have to take $a + b\sqrt{5}$ to $a + b\sqrt{5}$. That is, σ would be the neutral element ι. Because we are assuming it is not, the only choice is for σ to take $\sqrt{5}$ to $-\sqrt{5}$. Then, it must take $a + b\sqrt{5}$ to $a - b\sqrt{5}$, no matter what rational numbers a and b may be.

If you want a good algebra exercise, you can prove that if σ is the permutation of $\mathbf{Q}(f)$ defined by $\sigma(a + b\sqrt{5}) = a - b\sqrt{5}$, then σ *does* preserve the four arithmetic operations. (Remember, this means you have to check that $\sigma(u + v) = \sigma(u) + \sigma(v)$ and $\sigma(uv) = \sigma(u)\sigma(v)$ for all elements u and v in $\mathbf{Q}(f)$. The other two requirements,

$\sigma(u-v)=\sigma(u)-\sigma(v)$ and $\sigma(u/v)=\sigma(u)/\sigma(v)$, can be shown to follow from these two.)

So we see that the only two elements of $G(f)$ are ι and σ. In fact, we can think of $G(f)$ as the permutations of the two numbers $\sqrt{5}$ and $-\sqrt{5}$. The trivial permutation is given by ι, and the switching of the two is given by σ. So $G(f)$ is $\Sigma_{\{\sqrt{5},-\sqrt{5}\}}$.

EXAMPLE: $f(x)=x^3-5$. As soon as we go beyond degree 2, things get much more difficult and we can only tell you the answers. For ease of writing, we assign letters to the three roots of $f(x)$: $r=\sqrt[3]{5}$, $s=\omega\sqrt[3]{5}$, and $t=\omega^2\sqrt[3]{5}$—so they are the three complex numbers that give 0 when you plug them into $f(x)$. Here, ω is a complex number whose cube happens to be equal to 1. If you want to know, $\omega=-\frac{1}{2}+\frac{i\sqrt{3}}{2}$. And remember, every real number, such as $\sqrt[3]{5}$, is also considered to be a complex number, because it is an element of $\mathbf{C}$.

As we explained a few paragraphs ago, any element of $G(f)$ is determined by how it permutes the roots of f, or in this case how it permutes r, s, and t.

Now, which permutations are allowable depends on what $\mathbf{Q}(f)$ is, and that can be rather difficult to figure out. Certainly, $\mathbf{Q}(f)$ contains all numbers of the form $a+br+cr^2$, where a, b, and c are any rational numbers. So it becomes important to figure out whether s and t are also of that form. Now because $r=\sqrt[3]{5}$ is a real number, so is r^2 and so is every number of the form $a+br+cr^2$, with a, b, and c rational (or even real.)

Now, neither s nor t can be written in that form, because they are not real. It turns out that this gives you a lot of freedom. In fact, you can send r to either s or t and flesh that out into an element of $G(f)$. In fact, $G(f)$ turns out to be the whole group $\Sigma_{\{r,s,t\}}$, with six elements. All permutations of r, s, and t are possible results of applying an element of $G(f)$.

EXAMPLE: (This appears in an actual research paper by one of us (Ash et al., 1991, page 762).) Let $f(x)=x^3+x^2-20x-9$.

We call the three roots of this cubic u, v, and w. They are all irrational real numbers:

$$u = -4.786\ldots$$
$$v = -0.446\ldots$$
$$w = 4.230\ldots$$

In this case, it turns out that $G(f)$ has only three elements: ι, σ, and τ. As usual, ι is the neutral permutation, while σ causes the permutation $u \to v \to w \to u$ on the roots of f, and τ causes the permutation $u \to w \to v \to u$.

You may think that $G(f)$ should have six elements, because there are six possible permutations of the three numbers u, v, and w. However, each of the other three permutations (e.g., $u \to v \to u$, $w \to w$) fails to belong to the Galois club: They do not extend to permutations of $\mathbf{Q}(f)$ that preserve addition and multiplication.

That is enough examples. This is already looking too much like a textbook on Galois theory. You can look at some actual texts (Artin, 1998; Edwards, 1984; Fenrick, 1998; Gaal, 1998; Garling, 1986; Rotman, 1998; Stewart, 1989) if you want to see more complicated examples.

Digression: The Inverse Galois Problem

We say that two groups H and K are "isomorphic" if there is a one-to-one correspondence between them that is a morphism, that is, that preserves the group laws. For instance, in the examples above,

- $\Sigma_{\{\sqrt{5},-\sqrt{5}\}}$ is isomorphic to $\Sigma_{\{1,2\}}$,
- $\Sigma_{\{r,s,t\}}$ is isomorphic to $\Sigma_{\{1,2,3\}}$, and
- $G(f)$ in the last example is isomorphic to $\mathbf{F}_3$ with the group

law of addition, where

$$\iota \to 0$$
$$\sigma \to 1$$
$$\tau \to 2.$$

If H and K are isomorphic, then everything "grouplike" about them is the same. If H is infinite, K is infinite, and vice versa. Or if H has 43 elements, so does K and vice versa. If H is commutative, so is K and vice versa. And on and on, for all the fine structure you can think of, as long is you can state it in terms of the group law.

The inverse Galois problem is the following problem: Given any finite group H, can you find a **Z**-polynomial $f(x)$ with $G(f)$ isomorphic to H? If you can prove that this is always possible, give us a call! It is an important unsolved problem. It is not easy to see why it is so difficult, but you can take this as an indication of just how complicated these fields of the form $\mathbf{Q}(f)$ can be.

This problem may also give you a little insight into what mathematicians do. They do not just keep proving the same old theorems over and over, or add up large columns of numbers. Some of them work on the inverse Galois problem. This means trying to come up with a general proof where, given any H, you can always find an $f(x)$ as above. If that is too difficult, they try to do it for certain H's or classes of H's. Here there has been some success, and some very complicated and large finite groups are known to be isomorphic to Galois groups of the form $G(f)$. One of them is a huge group known as the "Monster." You will have to find a book on group theory to find out more about what that is. Suffice it to say that it has $2^{46} \cdot 3^{20} \cdot 5^{9} \cdot 7^{6} \cdot 11^{2} \cdot 13^{3} \cdot 17 \cdot 19 \cdot 23 \cdot 29 \cdot 31 \cdot 41 \cdot 47 \cdot 59 \cdot 71$ elements!

Two More Things

Here is a difficult and lengthy exercise:

EXERCISE: Let $f(x) = (x^2 - 2)(x^2 - 3)$. The roots of f are $\sqrt{2}$, $-\sqrt{2}$, $\sqrt{3}$, and $-\sqrt{3}$.

Show that $G(f)$ has exactly four elements, ι, σ, τ, and ρ, where ι is the identity permutation on the roots and the other permutations are the following:[3]

- $\sigma(\pm\sqrt{2}) = \mp\sqrt{2},\ \sigma(\pm\sqrt{3}) = \pm\sqrt{3};$
- $\tau(\pm\sqrt{2}) = \pm\sqrt{2},\ \tau(\pm\sqrt{3}) = \mp\sqrt{3};$
- $\rho(\pm\sqrt{2}) = \mp\sqrt{2},\ \rho(\pm\sqrt{3}) = \mp\sqrt{3}.$

SOLUTION: Here is a sketch: Let γ be an element of $G(f)$. We must show that γ is one of ι, σ, τ, or ρ. Now $\gamma(\sqrt{2})$ must be a root of $x^2 - 2$, so $\gamma(\sqrt{2}) = \sqrt{2}$ or $\gamma(\sqrt{2}) = -\sqrt{2}$. Similarly, $\gamma(\sqrt{3}) = \sqrt{3}$ or $\gamma(\sqrt{3}) = -\sqrt{3}$. Therefore, ι, σ, τ, and ρ are the only possibilities for γ.

The harder part is to show that all four possibilities work. First you have to identify $\mathbf{Q}(f)$, which you can check is the set of all numbers of the form $a + b\sqrt{2} + c\sqrt{3} + d\sqrt{6}$, where a, b, c, and d are rational numbers. Then check that each of ι, σ, τ, and ρ preserves addition and multiplication in $\mathbf{Q}(f)$.

To get you started, this is how you show that σ preserves addition: If $s = a + b\sqrt{2} + c\sqrt{3} + d\sqrt{6}$ and $s' = a' + b'\sqrt{2} + c'\sqrt{3} + d'\sqrt{6}$, then $\sigma(s) = a - b\sqrt{2} + c\sqrt{3} - d\sqrt{6}$ and $\sigma(s') = a' - b'\sqrt{2} + c'\sqrt{3} - d'\sqrt{6}$.

Now you check that $\sigma(s + s') = (a + a') - (b + b')\sqrt{2} + (c + c')\sqrt{3} - (d + d')\sqrt{6} = (a - b\sqrt{2} + c\sqrt{3} - d\sqrt{6}) + (a' - b'\sqrt{2} + c'\sqrt{3} - d'\sqrt{6}) = \sigma(s) + \sigma(s')$.

By the way, you may have noticed that in all our examples, the number of elements in $G(f)$ is the same as the number of rational parameters it takes to describe $\mathbf{Q}(f)$. This is always true, and is one of the main theorems of Galois Theory.

[3]The $\mp$, $\pm$ notation means that in any single equation, you can resolve $\pm$ as $+$ or $-$, in which case you must resolve $\mp$ as the opposite sign. For example, $\sigma(\pm\sqrt{2}) = \mp\sqrt{2}$ is a concise way of writing *two* equations: $\sigma(+\sqrt{2}) = -\sqrt{2}$ and $\sigma(-\sqrt{2}) = +\sqrt{2}$.

Chapter 14

THE RESTRICTION MORPHISM

Road Map

We patch together the Galois groups of polynomials into the absolute Galois group G via a type of morphism of groups called a "restriction morphism." This glues together all the $G(f)$'s and conversely gives us a method for studying G by looking at the $G(f)$'s. It also provides us with a lot of permutation representations of G.

If we somehow learn something about G, we can apply the restriction morphism to get information about a particular $G(f)$ in which we may be interested. Later, we will replace $G(f)$ with even more complicated objects related to **Z**-varieties. In this way, we can prove results about varieties such as Fermat's Last Theorem, or, more generally, theorems about Diophantine equations such as $x^p + y^q = z^r$.

The Big Picture and the Little Pictures

We have now considered several examples of Galois groups of single **Z**-polynomials. Mathematicians like to put things together into larger things whenever possible. This activity is related to the problem of understanding the universe. Are all the different parts of the universe completely different from one another, or are they connected by various relationships, especially by cause and effect? How strong are these connections? Can you go so far as to say that the whole universe, and every detail in it, is the expression of a "One," a single entity that somehow causes everything else? Unlike philosophers, mathematicians to some extent get to invent or arrange their universe the way they like. It is certainly useful to group similar things together. Sometimes, this causes one large object to form, which becomes a useful tool for looking at all the individual things. This is one way to look at G, the absolute Galois group of $\mathbf{Q}$.

What is the connection between G and the Galois group $G(f)$ of a given polynomial $f(x)$? There is a morphism from G to $G(f)$ called the *restriction morphism*. It is the crucial glue that holds all the possible $G(f)$'s together and welds them into G, or, conversely, that allows us to go from G, which is an infinite group, to the more understandable realm of the individual finite $G(f)$'s.

The world is nothing more than all the individuals and neighborhoods and countries that comprise it, fitted together with innumerable relationships of different kinds. In mathematics, the relationships are less varied and more mechanical. This difference makes mathematics easier, or at least more reliable, than politics. The basic relationship between G and $G(f)$, the restriction morphism, is a sort of passport that guarantees that $G(f)$ fits into G.

Here is how to describe the restriction morphism: Remember that $\mathbf{Q}^{\text{alg}}$ is the set of all those complex numbers that are the roots of $\mathbf{Z}$-polynomials. And G is the group of all one-to-one correspondences of $\mathbf{Q}^{\text{alg}}$ with itself that preserve all the arithmetic operations of addition, subtraction, multiplication, and division. What is $G(f)$? It consists of those permutations of the roots of $f(x)$ that preserve the arithmetic operations in the field of numbers $\mathbf{Q}(f)$ generated by those roots.

Now, as we saw in chapter 8, any element γ in G will permute the roots of any given $\mathbf{Z}$-polynomial. Of course, γ is very busy permuting *all* the roots of *all* the $\mathbf{Z}$-polynomials at the same time. But if we focus in on $f(x)$, we see that, in particular, γ is permuting *its* roots. Of course, because γ preserves all the arithmetic operations among all the algebraic numbers, in particular γ preserves the arithmetic operations in the field of numbers generated by the roots of $f(x)$. In other words, we can restrict our attention to see how γ acts on the roots of $f(x)$ and on the field they generate, forgetting about what γ may be doing to other algebraic numbers. When we do this, we see that γ has turned into an element of $G(f)$. In other words, there is an element σ of $G(f)$ that is specified by saying: σ does exactly what γ does in the field of numbers $\mathbf{Q}(f)$ generated by the roots of $f(x)$. This gives us a function from G to $G(f)$, which sends γ to σ.

Because the group law in both G and $G(f)$ is just composition of permutations, this function preserves the group law. In other words, it is a morphism. It is called the restriction morphism, and we will call it $r_{G(f)}$. If you start with a new **Z**-polynomial, say $h(x)$, and get a new Galois group $G(h)$ that acts in the field $\mathbf{Q}(h)$ generated by its roots, then we get another restriction morphism, and we will call it $r_{G(h)}$. And so on. Thus, we have infinitely many restriction morphisms, $r_{G(f)} : G \to G(f)$, $r_{G(h)} : G \to G(h)$, and so on.

Basic Facts about the Restriction Morphism

In this way, we obtain a lot of restriction morphisms. Every Galois group of a **Z**-polynomial is the target of exactly one of them. A big theorem in Field Theory tells us that every one of these restriction morphisms is "surjective."[1] This means that if we take any particular **Z**-polynomial f, and any allowable permutation σ of the roots of f (allowable in the sense that it is in $G(f)$), then we can find some element γ in G that does exactly the same permutation of the roots of this particular polynomial while doing who-knows-what to all of the other algebraic numbers—that is, $r_{G(f)}(\gamma) = \sigma$.

So G really does contain all the information of all the different $G(f)$'s. You get back information about $G(f)$ by applying $r_{G(f)}$ to whatever you might know about G. Therefore, G is like a Holy Grail. If we understood G completely, we would know a lot about all the different **Z**-polynomials and their Galois groups.

A bonus of wrapping all the information up into G is that certain theorems become easier to state or easier to remember. It helps to have the single object G as our focus. Also, we can

[1]This is the "existence theorem" that we mentioned back in chapter 8.

view each restriction morphism $r_{G(f)}: G \to G(f)$ as a permutation representation of G, because its target is a set of permutations (of the roots of f).

Let us think some more about the restriction morphism $r_{G(f)}$ from G to $G(f)$. The source is an infinite group and the target is a finite group. This means that necessarily, many different γ's will map to the same element σ of $G(f)$. In fact, for every element σ of $G(f)$, there will be infinitely many different elements γ of G such that $r_{G(f)}(\gamma) = \sigma$. What should we think about this? It is not difficult: If γ and γ' both map to σ, it just means that γ and γ' do the same things to the field generated by the roots of $f(x)$, but they may easily do different things to other fields, generated by roots of other **Z**-polynomials, so that overall they are different permutations in G.

But γ and γ' do not have complete freedom to differ. They have to do the same thing on any root, say b, of $f(x)$. So whatever they choose to do on some unrelated algebraic number, say a, they have to act consistently on $a + b$, $a - b$, ab, and a/b. Otherwise they will be thrown out of the G-club. It is not easy being in G and keeping track of all your commitments. It is exactly this intricate way in which all the different subfields of algebraic numbers fit together arithmetically that makes understanding G so difficult.

Another fact about the restriction morphisms: If you have an element γ in G, and you know about all its images $r_{G(f)}(\gamma)$, as $G(f)$ runs over all the Galois groups of all the **Z**-polynomials f, then you know γ uniquely. There is no more hidden information about γ. There is no Dr. Jekyll and Mr. Hyde phenomenon. Why? The personality and identity of γ are completely contained in how γ permutes the algebraic numbers $\mathbf{Q}^{\mathrm{alg}}$, and every algebraic number is a root of some f.

This is really important, so let us say it again in more detail. In contrast to the case of a single restriction morphism, if $r_{G(f)}(\gamma) = r_{G(f)}(\gamma')$ for *every* Galois group $G(f)$, then γ and γ' must really be the same element of G. Why? Because γ and γ' are the same if they do the same thing to every root of every **Z**-polynomial. So pick any **Z**-polynomial, say $h(x)$, and look at its Galois group $G(h)$. Because $r_{G(h)}(\gamma) = r_{G(h)}(\gamma')$, it follows that γ and γ' do the same thing to every root of $h(x)$. This holds for any **Z**-polynomial $h(x)$. So γ and γ' do the same thing to every root of every **Z**-polynomial. So γ and γ' are the same.

The importance of the restriction morphisms cannot be overstressed. The absolute Galois group is infinite, and we humans cannot grasp it all at once. All we see are its manifestations in our human world, namely, how it acts, via the restriction morphisms, on the various Galois groups of single **Z**-polynomials. At least theoretically, we can understand any single **Z**-polynomial if we put our minds to it. We cannot grasp G all at once. But by seeing what it does to the roots of individual **Z**-polynomials, we can hope to prove theorems about G.

In fact, we can sometimes do more. Sometimes there are whole infinite families of **Z**-polynomials we can study together. For example, there is the family $x^n - 1$, where $n = 1, 2, 3, \ldots$. We will look more closely at these in chapter 18 when we investigate one-dimensional Galois representations.

Examples

We finish this chapter with a couple of examples. The only elements in G that we can describe totally explicitly are the neutral element ι and complex conjugation c. So we will give one example for each.

EXAMPLE: Whatever **Z**-polynomial f may be, $r_{G(f)}(\iota)$ is always the identity permutation.

EXAMPLE: Refer to the example in the last chapter where $f(x) = x^3 - 5$ and use the notation introduced there for its three roots, r, s, t: $r = \sqrt[3]{5}$, $s = \omega\sqrt[3]{5}$, and $t = \omega^2\sqrt[3]{5}$. Now, $c(\sqrt[3]{5}) = \sqrt[3]{5}$ because $\sqrt[3]{5}$ is real. Also, $c(\omega) = \omega^2$, and $c(\omega^2) = \omega$ (which you can deduce from the fact that ω and ω^2 are the two roots of $x^2 + x + 1$, or you can work it out directly from the value of ω given in the previous chapter).

So you see that $r_{G(f)}(c)$ performs the permutation $r \to r$, $s \to t$, and $t \to s$.

Chapter 15

THE GREEKS HAD A NAME FOR IT

Road Map

There are two more major ingredients we need before we can pull everything together and explain how linear representations of Galois groups allow us to generalize quadratic reciprocity into a vast theory of generalized reciprocity laws (much of which is still only conjectural). Those two things are *characters* and *Frobenius elements*.

Characters are functions attached to linear representations of groups. That is one reason to use linear representations instead of permutation representations. In fact, characters are only one piece of the whole characteristic polynomial of a linear representation. Characters are sufficient for us in this book, although for a full exposition of generalized reciprocity laws we would need the whole characteristic polynomial.

Characters have no exclusive relationship to number theory. They can be defined for *any* matrix representation of *any* group. On the other hand, Frobenius elements belong to Galois groups and thus live only in a number-theoretic world.

We will explain characters in this chapter and Frobenius elements in the next chapter.

Traces

There are some amazing facts about finite groups and their linear representations. To explain them, we first need to talk about the *trace* of a square matrix.

DEFINITION: The *trace* of a square matrix is the sum of the diagonal elements.

The "diagonal elements" are those elements that go from the upper left-hand corner to the lower right-hand corner of a square matrix.

EXAMPLE: The trace of

$$\begin{bmatrix} 1 & 2 & 3 & 4 \\ 5 & 6 & 7 & 8 \\ 9 & 10 & 11 & 12 \\ 13 & 14 & 15 & 16 \end{bmatrix}$$

is $1 + 6 + 11 + 16 = 34$.

We apply this apparently innocuous definition to matrix representations of groups.

DEFINITION: For any matrix representation r of the finite group H, the *character* of an element g in H under r is the trace of $r(g)$. Often the character is written $\chi_r(g)$, where the Greek letter χ is used because it is the initial letter of the Greek word that means "character."

Here is a concrete example. Back on page 145, we defined a representation V of the permutation group $\Sigma_{\{1,2,3,4\}}$ which was a morphism of $\Sigma_{\{1,2,3,4\}}$ to $GL(4, \mathbf{Z})$, the group of 4-by-4 integer matrices. For instance, V takes the permutation σ, described by

$$1 \to 1$$

$$2 \to 3$$

$$3 \to 4$$

$$4 \to 2$$

to the matrix

$$M = \begin{bmatrix} 1 & 0 & 0 & 0 \\ 0 & 0 & 0 & 1 \\ 0 & 1 & 0 & 0 \\ 0 & 0 & 1 & 0 \end{bmatrix}.$$

Then $\chi_V(\sigma) = \text{trace}(M) = 1 + 0 + 0 + 0 = 1$.

In Greek, the word "character" denotes some outstanding feature of a thing that enables us to identify that thing. Here is the definition from Liddell and Scott's *Greek Lexicon* (school edition):

> that which is cut in or marked, the impress or stamp on coins, seals, etc.... metaphorically the mark or token impressed on a person or thing, a characteristic, distinctive mark, character... a likeness, image, exact representation.

It is the middle meaning that is relevant to us here, but notice how at the end of the definition, the meaning is tied up with the concept of representation!

Suppose that H is a group and r is an n-dimensional linear representation[1] of H over k, where k is some field. This means that r is a function from H to the group $\text{GL}(n,k)$, and that this function is a morphism (i.e., $r(g \circ h) = r(g)r(h)$). Then the character χ_r is a function from H to k. In symbols, if $r : H \to \text{GL}(n,k)$, then $\chi_r : H \to k$.

[1] Remember that "linear representation" is synonymous with "matrix representation." We retain both terms for the sake of elegant variation.

Conjugacy Classes

FACT: Any group H splits up into subsets C with the following two properties:
1. Each subset C is obtained in the following way: You take any group element, say x. Then you take *all* of the elements of the group, call them g's, and form the group products $g \circ x \circ g^{-1}$. Notice that x itself is one of these products, because $e \circ x \circ e^{-1} = x$ (where e, as usual, is the neutral element). The subset that consists of all these $g \circ x \circ g^{-1}$'s is one of the C's. For example, x is an element of the C that you get by starting with x. It does not matter which x in C you start with. You get the same bunch of elements, namely C!
2. Any two elements of C have the same character value under *every* representation r. In symbols: For any matrix representation of the group H, call it r, if x and y are in the same C, then $\chi_r(x) = \chi_r(y)$.

These subsets C are called the "conjugacy classes of H." The C's are often quite large. The larger they are, the more the group law of H fails to be commutative. At any rate, in a commutative group, each conjugacy class contains only one element. This is because if the group law is commutative, then $g \circ x \circ g^{-1} = g \circ g^{-1} \circ x = e \circ x = x$.

The definition of the C's is actually accomplished just by property 1. There is only one way of splitting up H into C's that possess property 1. The proof is just an exercise using the axioms of the group law. Then property 2 is not too difficult to prove if you know some linear algebra. The really amazing fact—and it is a fact about representation theory—is that if x is in C and y is in C', which is some *other* conjugacy class, then there is some matrix representation r such that $\chi_r(x) \neq \chi_r(y)$. We will give an example of this fact a little later for the group $\Sigma_{\{1,2,3\}}$.

Even more amazing is the important role that the characters play in the number theory of Galois groups. Before we get to that, though, we have some more to explain about group representations and their characters.

Examples of Characters

To illustrate these concepts further, it will be helpful to have a concrete example. Suppose $H = \Sigma_{\{1,2,3\}}$, the permutation group of $\{1, 2, 3\}$. We call it Σ_3 for short.

We need to refer to the elements of Σ_3 in terms of just two of those elements. We let σ be the following permutation:

$$1 \to 2$$

$$2 \to 3$$

$$3 \to 1.$$

Let τ be this permutation:

$$1 \to 2$$

$$2 \to 1$$

$$3 \to 3.$$

It is a fact that all of the elements in Σ_3 can be written in terms of σ and τ (and the identity permutation e). We usually use power notation for the composition of a permutation with itself, and so σ^2 is another way to write $\sigma \circ \sigma$, which is the permutation

$$1 \to 3$$

$$2 \to 1$$

$$3 \to 2.$$

The six elements of Σ_3 are $\{e, \sigma, \sigma^2, \tau, \tau \circ \sigma, \tau \circ \sigma^2\}$.

EXERCISE: Show that $\sigma^3 = \tau^2 = e$, and that $\sigma \circ \tau = \tau \circ \sigma^2$.

EXERCISE: Show that Σ_3 has three conjugacy classes: $\{e\}$, $\{\sigma, \sigma^2\}$, and $\{\tau, \tau \circ \sigma, \tau \circ \sigma^2\}$.

SOLUTION: Recall that if x and y are in the same conjugacy class, then there is some element g in the group so that $y = g \circ x \circ g^{-1}$. Now, it is not too difficult to check that

$g \circ e \circ g^{-1}$ will always be e, so e is the only element in its conjugacy class. If you do a bit of computation, you can check that $\tau \circ \sigma \circ \tau^{-1} = \sigma^2$, so σ and σ^2 are in the same conjugacy class. A bit more work gives the two equations $\sigma \circ \tau \circ \sigma^{-1} = \tau \circ \sigma$ and $\sigma^2 \circ \tau \circ (\sigma^2)^{-1} = \tau \circ \sigma^2$, so τ, $\tau \circ \sigma$, and $\tau \circ \sigma^2$ are in the same conjugacy class. And then more trial-and-error shows that there is no element g in Σ_3 such that $g \circ \sigma \circ g^{-1} = \tau$, and similarly there is no way to conjugate any element in the second conjugacy class and get something in the third one.

In order to define a representation r, we need to say what $r(\sigma)$ and $r(\tau)$ are. This will totally determine r, because every element of Σ_3 can be written in terms of σ and τ. Because r is a representation, we must have $r(\tau \circ \sigma) = r(\tau)r(\sigma)$, and $r(\sigma^2) = r(\sigma)^2$, and so on.

One representation of Σ_3 in $\mathrm{GL}(3, \mathbf{Z})$ is given by

$$r(\sigma) = \begin{bmatrix} 0 & 1 & 0 \\ 0 & 0 & 1 \\ 1 & 0 & 0 \end{bmatrix}, \qquad r(\tau) = \begin{bmatrix} 0 & 1 & 0 \\ 1 & 0 & 0 \\ 0 & 0 & 1 \end{bmatrix}.$$

But just choosing matrices for $r(\sigma)$ and $r(\tau)$ does not guarantee that r really extends to a representation. Remember that $r(e)$ must be the identity matrix, I, and this means that there are a few things to check before we know that this is indeed a representation:

EXERCISE: Check that $r(\sigma)^3 = I$ and $r(\tau)^2 = I$.

In fact, there is yet another relationship that these matrices must satisfy. Because $\sigma \circ \tau = \tau \circ \sigma^2$, if r is going to be a representation, we must have $r(\sigma)r(\tau) = r(\tau)r(\sigma)^2$.

EXERCISE: Check that the matrix equation $r(\sigma)r(\tau) = r(\tau)r(\sigma)^2$ is true.

It can be proven that this is enough checking for this group Σ_3 and in fact r does extend to a bona fide representation from Σ_3 to $\mathrm{GL}(3, \mathbf{Z})$.

EXERCISE: Check that the representation r is faithful.

SOLUTION: Remember that a representation is faithful if the *only* element in the source that goes to the neutral element of the target is the neutral element of the source. In our situation, we need to see that if x is a permutation in Σ_3, and $r(x) = I$, then $x = e$.

The easiest way to check that this representation is faithful is to list $r(x)$ for all x in Σ_3 other than the identity, and see that $r(x)$ is never I. We see above that $r(\sigma)$ is not the identity and $r(\tau)$ is not the identity. Next, we compute

$$r(\sigma^2) = r(\sigma)^2 = \begin{bmatrix} 0 & 1 & 0 \\ 0 & 0 & 1 \\ 1 & 0 & 0 \end{bmatrix}^2 = \begin{bmatrix} 0 & 0 & 1 \\ 1 & 0 & 0 \\ 0 & 1 & 0 \end{bmatrix}.$$

Next, we check

$$r(\tau \circ \sigma) = r(\tau)r(\sigma) = \begin{bmatrix} 0 & 1 & 0 \\ 1 & 0 & 0 \\ 0 & 0 & 1 \end{bmatrix} \begin{bmatrix} 0 & 1 & 0 \\ 0 & 0 & 1 \\ 1 & 0 & 0 \end{bmatrix} = \begin{bmatrix} 0 & 0 & 1 \\ 0 & 1 & 0 \\ 1 & 0 & 0 \end{bmatrix}.$$

Finally, we check

$$r(\tau \circ \sigma^2) = r(\tau)r(\sigma)^2 = \begin{bmatrix} 0 & 1 & 0 \\ 1 & 0 & 0 \\ 0 & 0 & 1 \end{bmatrix} \begin{bmatrix} 0 & 0 & 1 \\ 1 & 0 & 0 \\ 0 & 1 & 0 \end{bmatrix} = \begin{bmatrix} 1 & 0 & 0 \\ 0 & 0 & 1 \\ 0 & 1 & 0 \end{bmatrix}.$$

Because none of these products give I as the answer, we know that the only permutation solving $r(x) = I$ is the neutral one, and therefore r is a faithful representation.

Now we write down the character of r. Because we have said that $\chi_r(x) = \chi_r(y)$ whenever x and y are in the same conjugacy class, we only have to write down the value of χ_r on one volunteer from each conjugacy class, say e, σ, and τ. By computing the traces of the matrices $r(e)$, $r(\sigma)$ and $r(\tau)$, you can check that $\chi_r(e) = 3$, $\chi_r(\sigma) = 0$, and $\chi_r(\tau) = 1$.

You may want to check that the character is really the same on all elements of the same conjugacy class. Here is a complete list of the values of the character: $\chi_r(\sigma) = 0$, $\chi_r(\sigma^2) = 0$, $\chi_r(\tau) = 1$, $\chi_r(\tau \circ \sigma) = 1$, $\chi_r(\tau \circ \sigma^2) = 1$, and $\chi_r(e) = 3$. We saw that σ and σ^2 are in the same conjugacy class, and, sure enough, $\chi_r(\sigma) = \chi_r(\sigma^2)$. What about the other conjugacy classes? The neutral element e is in a class with no other elements, so there is nothing to check. As for the third conjugacy class, it contains τ, $\tau \circ \sigma$, and $\tau \circ \sigma^2$, and we do indeed have $\chi_r(\tau) = \chi_r(\tau \circ \sigma) = \chi_r(\tau \circ \sigma^2)$.

Here is another representation of Σ_3, this time to $\mathrm{GL}(1, \mathbf{Z})$. We will call it sgn, which is short for "sign." Furthermore, because $\mathrm{GL}(1, \mathbf{Z})$ consists of 1-by-1 matrices, we eliminate the brackets and just write these matrices as numbers. We define $\operatorname{sgn}(\sigma) = 1$, and $\operatorname{sgn}(\tau) = -1$. It is not obvious that this defines a representation, but you can check that it satisfies the requirements to do so.

EXERCISE: Check that $\operatorname{sgn}(\sigma)^3 = 1$, $\operatorname{sgn}(\tau)^2 = 1$, and $\operatorname{sgn}(\sigma)\operatorname{sgn}(\tau) = \operatorname{sgn}(\tau)\operatorname{sgn}(\sigma^2)$.

EXERCISE: Show that this is not a faithful representation.

SOLUTION: Remember that showing that a representation is faithful means showing that the only element of the source group that gets mapped to the neutral element is the neutral element. But we just wrote down the equation $\operatorname{sgn}(\sigma) = 1$, so we can see straight from the definition of the representation sgn that it is not faithful.

We now write down the character of sgn. For sgn, things are very simple, because χ_{sgn} is just sgn; the trace of a 1-by-1 matrix is just the matrix itself, considered as a number. What about conjugacy classes? Even though σ and e are in different conjugacy classes, we see that $\chi_{\text{sgn}}(e) = \chi_{\text{sgn}}(\sigma) = \chi_{\text{sgn}}(\sigma^2) = 1$, and $\chi_{\text{sgn}}(\tau) = \chi_{\text{sgn}}(\tau \circ \sigma) = \chi_{\text{sgn}}(\tau \circ \sigma^2) = -1$.

Comparing the character of r to the character of sgn, you can see the truth of our assertion that if $C \neq C'$ are two different conjugacy classes, you can find two representations that take different values on them. Incidentally, for most groups, it takes more than just two representations to separate all possible pairs of conjugacy classes.

Every group has a "trivial representation" which by definition is the morphism that sends every element of the group to the neutral element of the target. For example, Σ_3 has the trivial representation id to GL(1, **Z**) which takes every element to 1. Its trace is easy to write down: $\chi_{\text{id}}(e) = \chi_{\text{id}}(\sigma) = \chi_{\text{id}}(\tau) = 1$.

Another representation of Σ_3 can be defined into the group GL(2, $\mathbf{F}_2$). We call this representation R, and define

$$R(\sigma) = \begin{bmatrix} 0 & 1 \\ 1 & 1 \end{bmatrix}, \qquad R(\tau) = \begin{bmatrix} 0 & 1 \\ 1 & 0 \end{bmatrix}.$$

Again, this is not obviously a representation:

EXERCISE: Check that $R(\sigma)^3 = 1$, $R(\tau)^2 = 1$, and $R(\sigma)R(\tau) = R(\tau)R(\sigma^2)$.

SOLUTION: Do not forget that in this computation, the entries in the matrices are from $\mathbf{F}_2$, so that we use the equation $1 + 1 = 0$.

To begin with,

$$R(\sigma^2) = \begin{bmatrix} 0 & 1 \\ 1 & 1 \end{bmatrix}^2 = \begin{bmatrix} 1 & 1 \\ 1 & 0 \end{bmatrix}.$$

We can also compute that

$$R(\tau \circ \sigma) = \begin{bmatrix} 0 & 1 \\ 1 & 0 \end{bmatrix} \begin{bmatrix} 0 & 1 \\ 1 & 1 \end{bmatrix} = \begin{bmatrix} 1 & 1 \\ 0 & 1 \end{bmatrix},$$

$$R(\tau \circ \sigma^2) = \begin{bmatrix} 0 & 1 \\ 1 & 0 \end{bmatrix} \begin{bmatrix} 0 & 1 \\ 1 & 1 \end{bmatrix}^2 = \begin{bmatrix} 1 & 0 \\ 1 & 1 \end{bmatrix}.$$

We will leave the rest of the computations to you.

EXERCISE: Show that R is a faithful representation.

What is the character of the representation R? Here we must remember that χ_R will be an element of $\mathbf{F}_2$, not of $\mathbf{Z}$. So we get $\chi_R(\sigma) = 1$, $\chi_R(\tau) = 0$, and (here is the surprise) $\chi_R(e) = 1 + 1 = 0$. Again, we can check that χ_R is constant for all members of a conjugacy class: We see that $\chi_R(\tau) = \chi_R(\tau \circ \sigma) = \chi_R(\tau \circ \sigma^2)$, and $\chi_R(\sigma) = \chi_R(\sigma^2)$.

How the Character of a Representation Determines the Representation

The amazing thing is that the character χ_r of a representation r *determines* r in an appropriate sense, so it is in fact a distinct token of the representation. The character is sometimes easier to think about than r itself, because it assigns a single number to each element of the source group, rather than a whole matrix.

To see in what sense a character determines a representation, perform the following experiment. Write down any 3-by-3 matrix A and find its trace. Now pick your favorite invertible 3-by-3 matrix B and compute BAB^{-1}. You will see that the trace of the matrix BAB^{-1} is the same as the trace of A! Here is an example, but you should do your own. We will let

$$B = \begin{bmatrix} 3 & 10 & 65 \\ 4 & 14 & 91 \\ 0 & 1 & 7 \end{bmatrix},$$

and then we can compute that

$$B^{-1} = \begin{bmatrix} 7 & -5 & 0 \\ -28 & 21 & -13 \\ 4 & -3 & 2 \end{bmatrix}.$$

You can check that $BB^{-1} = B^{-1}B = I$. We pick an A at random:

$$A = \begin{bmatrix} 2 & 3 & 5 \\ 7 & 11 & 13 \\ 17 & 19 & 23 \end{bmatrix}.$$

It is easy to see that the trace of A is $2 + 11 + 23 = 36$. Finally, we have

$$BAB^{-1} = \begin{bmatrix} 3 & 10 & 65 \\ 4 & 14 & 91 \\ 0 & 1 & 7 \end{bmatrix} \begin{bmatrix} 2 & 3 & 5 \\ 7 & 11 & 13 \\ 17 & 19 & 23 \end{bmatrix} \begin{bmatrix} 7 & -5 & 0 \\ -28 & 21 & -13 \\ 4 & -3 & 2 \end{bmatrix}$$

$$= \begin{bmatrix} -23{,}085 & 17{,}609 & -14{,}322 \\ -32{,}309 & 24{,}645 & -20{,}045 \\ -2{,}454 & 1{,}872 & -1{,}524 \end{bmatrix},$$

and sure enough the trace of BAB^{-1} is $-23{,}085 + 24{,}645 - 1{,}524 = 36$. This is the reason that elements in a single conjugacy class have the same character as one another, whatever representation r you choose. Namely, if x and y are in the same conjugacy class, then for some z we have the equation $z \circ x \circ z^{-1} = y$, so that $r(z) \circ r(x) \circ r(z)^{-1} = r(y)$, so $r(x)$ and $r(y)$ have the same trace.

So if r is our representation, then $Br(g)B^{-1}$ will have the same trace as $r(g)$.[2] We now ask the question: To what extent can we reconstruct the representation if all we know is its character? It looks as if the character of r could hardly determine r. Just change r by picking an invertible matrix B_g for each g, replace $r(g)$ by $B_g r(g) B_g^{-1}$, and the traces, and hence the character, will stay the same.

But wait a minute. Why should the function $g \to B_g r(g) B_g^{-1}$ be a morphism? In general, it will not be.

However, if we choose B_g to be a *constant* invertible matrix B, and define $s(g) = Br(g)B^{-1}$ for every g in H, then you can check that s is itself a morphism.

EXERCISE: If $r(g)$ is a linear representation of a group H (so that $r(g \circ h) = r(g)r(h)$), and B is a constant invertible matrix

[2]Note that $Br(g)B^{-1}$ is the product of *three* matrices: B, $r(g)$, and B^{-1}.

of the same size with entries from the same number system, and $s(g) = Br(g)B^{-1}$, show that $s(g \circ h) = s(g)s(h)$.

SOLUTION: We write it out:

$$\begin{aligned} s(g)s(h) &= (Br(g)B^{-1})(Br(h)B^{-1}) = (Br(g))(B^{-1}B)(r(h)B^{-1}) \\ &= Br(g)r(h)B^{-1} \end{aligned}$$

because $BB^{-1} = I$. And the last term equals $Br(g \circ h)B^{-1}$ because r is a morphism, and in turn $Br(g \circ h)B^{-1} = s(g \circ h)$.

In summary, $s(g \circ h) = s(g)s(h)$ for any g and h in H, so s is a morphism—in fact, it is another linear representation of H.

Now, the representation s is essentially the same as r—it is just carrying around a B in front and a B^{-1} in back. If you know linear algebra, then you know that s represents the same geometry as r, except with respect to a different basis. If you change your coordinate axes, you will change the numbers that represent events, but you do not change anything in the reality itself. Similarly, if you replace r by s, you are making a superficial change in your description, but the underlying representation is essentially the same. Of course, it is not *exactly* the same, so we need a new word. We say that r and s are *equivalent* representations.

Take our example of r from above. Let $s(g) = Br(g)B^{-1}$, where B is the matrix we used above. Then we can compute that

$$\begin{aligned} s(\sigma) = Br(\sigma)B^{-1} &= \begin{bmatrix} 3 & 10 & 65 \\ 4 & 14 & 91 \\ 0 & 1 & 7 \end{bmatrix} \begin{bmatrix} 0 & 1 & 0 \\ 0 & 0 & 1 \\ 1 & 0 & 0 \end{bmatrix} \begin{bmatrix} 7 & -5 & 0 \\ -28 & 21 & -13 \\ 4 & -3 & 2 \end{bmatrix} \\ &= \begin{bmatrix} 411 & -292 & -19 \\ 581 & -413 & -24 \\ 53 & -38 & 2 \end{bmatrix} \end{aligned}$$

and

$$s(\tau) = Br(\tau)B^{-1} = \begin{bmatrix} 3 & 10 & 65 \\ 4 & 14 & 91 \\ 0 & 1 & 7 \end{bmatrix} \begin{bmatrix} 0 & 1 & 0 \\ 1 & 0 & 0 \\ 0 & 0 & 1 \end{bmatrix} \begin{bmatrix} 7 & -5 & 0 \\ -28 & 21 & -13 \\ 4 & -3 & 2 \end{bmatrix}$$

$$= \begin{bmatrix} 246 & -182 & 91 \\ 350 & -259 & 130 \\ 35 & -26 & 14 \end{bmatrix}.$$

And, sure enough, we discover that $\chi_s(\sigma) = \chi_r(\sigma)$, $\chi_s(\tau) = \chi_r(\tau)$, and similarly for the other four elements of Σ_3. Even though the matrices for s look much more complicated than those for r, the representations r and s are essentially the same.

EXERCISE: Check that $\chi_r(\tau \circ \sigma) = \chi_s(\tau \circ \sigma)$.

Here is the amazing theorem:

THEOREM 15.1: Let r and s be two linear representations of a finite group into $\mathrm{GL}(n, \mathbf{C})$. Then $\chi_r = \chi_s$ if and only if r and s are equivalent.

So the character of a linear representation has been aptly named.

The existence of the character and Theorem 15.1 are two of the main reasons that linear representations are so useful. Unfortunately, it is beyond the scope of this book even to attempt to sketch a proof of this amazing theorem. Why is it amazing? A character is a lot simpler than a representation. The character values $\chi_r(g)$ are just single numbers—not big matrices—and they are constant on conjugacy classes. Yet equivalent representations have the *same* character and nonequivalent ones have *different* characters (at least when we are using matrix representations with matrices containing *complex* numbers).

If the representations r and s take values in $\mathrm{GL}(n, \mathbf{F}_p)$ rather than $\mathrm{GL}(n, \mathbf{C})$, then the corresponding theorem is somewhat more complicated and we will omit it. Of course, equivalent representations always have the same character.

Prelude to the Next Chapter

Group representations and their characters were first studied over 100 years ago. It was seen immediately that characters were very important for understanding groups and their representations. But in number theory, the character takes on a new life in the world of reciprocity laws. If the representation is of a Galois group $G(f)$ of a field $\mathbf{Q}(f)$, then the character can tell us a lot about the field $\mathbf{Q}(f)$, and about how the polynomial $f(x)$ behaves modulo various primes p. This is especially true if the representation is a faithful one.

To repeat: If r is a linear representation of a finite Galois group $G(f)$, then the character of r reveals many important and fascinating number-theoretic properties of f and $G(f)$. How? We have to apply it to special elements of $G(f)$, called *Frobenius elements*. We will define them in the next chapter.

Before going on to the next chapter, though, remember the restriction morphism of the absolute Galois group G. If we have a representation r of a finite Galois group $H = G(f)$, we can always compose it (as a function) with the restriction morphism $r_H : G \to H$ to get a representation of G which has basically the same information as r. We like to do this, because then we can refer everything to the same big group G. It will be easiest for us to discuss the Frobenius elements first as elements in G and then restrict them (via r_H) to H, and then we can apply r.

Digression: A Fact about Rotations of the Sphere

In a footnote on page 15, we mentioned something that may have been on your mind ever since. We had an example of two elements g and h in SO(3) and we showed that $g \circ h \neq h \circ g$. And then we mentioned that in fact $g \circ h$ and $h \circ g$ will be rotations through the same number of degrees, just with different axes of rotation.

This can be proved by using traces. For simplicity of notation, we think of each rotation in SO(3) as being *equal* to the matrix in $\mathrm{GL}(3, \mathbf{R})$ that represents it, as described on page 145. The key is that $g \circ h$ and $h \circ g$ are in the same conjugacy class, because $g \circ h = h^{-1} \circ (h \circ g) \circ h$. Therefore, the trace of $g \circ h$ is the same as the trace of $h \circ g$. Now, a bit of trigonometry will show you that if A is a rotation of $\theta°$ in SO(3), then the trace of A is $1 + 2\cos\theta°$. Therefore, if two different elements in SO(3) have the same trace, then they must be rotations by the same amount.

Chapter 16

FROBENIUS

Road Map

Here we explain how to describe some very important elements of the absolute Galois group G. For each prime number p there is a set of elements of G called Frobenius elements at p. As usual in mathematics, they are important partly because we can describe them with some explicitness, and partly because we can do something interesting with them. What we will do with them in this book is to make generalized reciprocity laws, beginning in the next chapter.

Frobenius elements are difficult to define precisely. We give an explanation in this chapter of a working definition. The completely correct characterization is presented in a (probably opaque) appendix to this chapter, so you can see what it is like. We also discuss "ramification" in this chapter; another appendix gives added details. These appendices briefly define many concepts whose detailed explication would form a large chunk of a text on algebraic number theory.

Something for Nothing

Ferdinand Georg Frobenius was the nineteenth-century German mathematician who invented the method of using characters to study group representations.[1] Today, his name is used for (among other things) particular elements in the absolute Galois group G, and also their restriction to the Galois groups of various polynomials, that is, for particular permutations of the roots of various **Z**-polynomials.

Whenever you can get something for nothing in mathematics, you take it. Of course, "nothing" is a relative term. What we mean in this case is that we can define certain elements of every Galois group by using an easily applied *general* theorem that tells us significant information about them. These are the Frobenius elements.

To take a simpler example first, let L be the field $\mathbf{Q}(f)$ for some **Z**-polynomial f. We know that L is composed of certain complex numbers. Let τ be complex conjugation.[2] Then τ defines an element of the Galois group $G(f)$ of L: If $x + iy$ is any number in L, $\tau(x + iy)$ is again a number in L, and τ respects addition, subtraction, multiplication, and division. Also, we have a formula for it: $\tau(x + iy) = x - iy$. Thus, τ is always there, and it is an important element of the Galois group $G(f)$. It is the neutral element if and only if L actually contains only real numbers, because then L does not contain any numbers of the form $x + iy$ with $y \neq 0$. We can make τ into an honorary Frobenius element at infinity, and call it Frob_∞. This is purely a notational convention, and does not help us to define Frob_p when p is a prime. However, this notation Frob_∞ is common in the research literature.

The other Frobenius elements are at p for each prime p, and we now begin to define them. If you choose to skip the rest of this discussion, you need to know only:

- Frob_p is a particular element of G.
- Actually, this is a lie, because we cannot define Frob_p so precisely. Really, given p, there is defined a certain conjugacy class in G (depending on p), and Frob_p is taken

---

[1] In fact, Frobenius deserves credit for inventing group representations themselves.

[2] We have used the letter τ before to stand for various things, but here we will use it for the specific function of complex conjugation. We have been using c for complex conjugation, but from now on we prefer Greek letters for elements of Galois groups.

to be any element in that conjugacy class. (We briefly explained what a conjugacy class is in the previous chapter.)

- In fact, we just lied again: Frob_p really is a union of conjugacy classes. To be precise, for any p, we define a set $\mathscr{F}(p)$, which is a union of conjugacy classes inside of the absolute Galois group G. We let Frob_p refer to any element of $\mathscr{F}(p)$. Because there is a choice involved here, we have to be careful when we talk about Frob_p as if it were a single element of G.

Now suppose that r is a matrix representation of the absolute Galois group G, or a Galois representation, for short. The idea is that we are never going to have a naked Frob_p in any formula. We will always be doing something to Frob_p first, of which the result will be the same no matter which element of $\mathscr{F}(p)$ we take as a stand-in. We are going to discuss not $r(\mathrm{Frob}_p)$, which might depend on the particular element of the set $\mathscr{F}(p)$ we chose, but $\chi_r(\mathrm{Frob}_p)$. Because the character of a representation of a group is constant on all elements of a conjugacy class, the fact that we have a choice about Frob_p will not matter when we end up talking only about $\chi_r(\mathrm{Frob}_p)$.

Good Prime, Bad Prime

Because Frob_p is not just a single conjugacy class but a whole bunch of them, it is not true that $\chi_r(\mathrm{Frob}_p)$ is always well-defined (i.e., independent of choices). We will always have to make an assumption about the relationship between r and p that will eliminate any ambiguity.

There is a concept called "ramification" that we will try to explain later in this chapter. Again, if you want to skip the explanation, what you need to know is that every Galois representation r comes with a set of ramified primes. We usually call this set S. (Sometimes mathematicians are not that imaginative.) If p is *not* in the set S, then the character value $\chi_r(\mathrm{Frob}_p)$ is well-defined. This means that no matter what element σ is chosen from $\mathscr{F}(p)$, $\chi_r(\sigma)$ will always be the same number. The great thing is that in our theorems we will only need to refer to this character value $\chi_r(\mathrm{Frob}_p)$, and not to the whole matrix $r(\mathrm{Frob}_p)$, which is not well-defined.

If p is in the set S, then $\chi_r(\mathrm{Frob}_p)$ is *not* well-defined, and we do not discuss Frob_p in relationship to r. We cannot. Frustrated, we often refer to S as the set of "bad" primes for the representation r. We are not being judgmental. We are just projecting our frustration onto the primes.

In most of the examples and theorems that are important in number theory, the set S of bad primes is finite. Usually, knowledge of S comes if we have knowledge of how a particular representation r is constructed. You will see examples of this later.

We should be decorating the set of bad primes S with a subscript, such as S_r, because S depends on what representation r we are studying. But that just gives us more to worry about, and because we will only deal with one r at a time, we just leave it out.

In summary, any Galois representation r comes with a set S of bad primes, and if p is not in S, then $\chi_r(\mathrm{Frob}_p)$ is a dandy, well-defined, unambiguous number which we can use in our formulas.

For completeness, we are going to tell you more about how to find Frob_p, but you could skim this material during your first reading of the book.

Algebraic Integers, Discriminants, and Norms

We start by telling you about parts of algebraic number theory that are not as complicated as they might seem at first:

> **DEFINITION**: An *algebraic integer* is an element of $\mathbf{Q}^{\mathrm{alg}}$ that is a root of a $\mathbf{Z}$-polynomial that has a leading coefficient[3] of 1. The set of all algebraic integers is written $\overline{\mathbf{Z}}$. If α is an

[3]The *leading coefficient* of a polynomial is the coefficient of the highest power of x in the polynomial. For example, the leading coefficient of the polynomial $3x^5 - 2x^2 + 7x - 1$ is 3.

element of $\overline{\mathbf{Z}}$, then the **Z**-polynomial of smallest degree with first coefficient 1 which has α as a root is called the *minimal polynomial* for α. *Note*: Most algebraic integers are *not* ordinary integers.

It is easy to find an element of $\overline{\mathbf{Z}}$: Find a **Z**-polynomial whose first coefficient is 1. For example, any solution of $x^5 + 41x^4 + 32x + 11 = 0$ is an algebraic integer. (Of course, we write x^5 instead of the longer $1x^5$, but the coefficient of x^5 is still 1.) It is less obvious—in fact, rather difficult to show—that sums and products of algebraic integers are algebraic integers. Quotients of algebraic integers are not necessarily algebraic integers. The reason for the terminology "algebraic integer" is that these are the elements of $\mathbf{Q}^{\text{alg}}$ that mimic ordinary integers: They can be added and multiplied, but not divided, without leaving the realm of algebraic integers. Of course you can divide them, but the answer may well be only an algebraic number, not an algebraic *integer* any more. Every element of $\mathbf{Q}^{\text{alg}}$ is a quotient of elements of $\overline{\mathbf{Z}}$. And one more important fact about $\overline{\mathbf{Z}}$: An ordinary integer is always an algebraic integer (e.g., 137 solves $x^1 - 137 = 0$), and, conversely, the *only* rational numbers that are algebraic integers are the ordinary integers. You can symbolize this last sentence by writing $\overline{\mathbf{Z}} \cap \mathbf{Q} = \mathbf{Z}$.

Next, we take a bit of a detour to discuss the *discriminant* of a polynomial. You may remember this term from high-school algebra: The discriminant of the quadratic polynomial $ax^2 + bx + c$ is $b^2 - 4ac$. You may even remember that the discriminant of a quadratic polynomial tells you if the polynomial has 0, 1, or 2 real roots. This, of course, is another way of saying how many points there are on the variety $S(\mathbf{R})$ associated with the polynomial. So it is not surprising that there is a generalization of the discriminant to higher degrees that will prove useful to us.

DEFINITION: Suppose that $f(x) = x^n + c_{n-1}x^{n-1} + \cdots + c_1x + c_0$ is a **Z**-polynomial with roots $\alpha_1, \alpha_2, \ldots, \alpha_n$, so that $f(x) = (x - \alpha_1)(x - \alpha_2)\cdots(x - \alpha_n)$. (Notice that because we are assuming that $f(x)$ starts with the coefficient 1, all of these roots are in $\overline{\mathbf{Z}}$.) The *discriminant* of the polynomial, which we will write Δ_f, is defined by

$$\Delta_f = \prod_{1 \le i < j \le n} (\alpha_i - \alpha_j)^2.$$

The right-hand side of this formula says to take all of the roots of the polynomial $f(x)$, compute all of the differences between all pairs of roots, square these differences, and multiply these numbers together to get the discriminant.

It is far from obvious, but Δ_f is an ordinary integer, that is, an element of $\mathbf{Z}$, not just an algebraic integer. So, one way to compute it is to compute all of the roots numerically to a lot of decimal places, compute the product in the equation, and then round it off to the nearest integer. For example, if we start with the polynomial $x^4 - 13x^3 + 22x^2 + 19x + 11$, a good computer can find that the four roots are approximately 10.78887214722256, 2.9556460085645, $-0.37225907789 - 0.454290204168i$, and $-0.37225907789 + 0.454290204168i$. We can now compute the product of the squares of the differences:

$$\begin{aligned}
&(10.78887214722256 - 2.9556460085645)^2 \\
&\quad \times \quad (10.78887214722256 - (-0.37225907789 - 0.454290204168i))^2 \\
&\quad \times \quad (10.78887214722256 - (-0.37225907789 + 0.454290204168i))^2 \\
&\quad \times \quad (2.9556460085645 - (-0.37225907789 - 0.454290204168i))^2 \\
&\quad \times \quad (2.9556460085645 - (-0.37225907789 + 0.454290204168i))^2 \\
&\quad \times \quad ((-0.37225907789 - 0.454290204168i) \\
&\qquad\quad - (-0.37225907789 + 0.454290204168i))^2 \\
&\quad = \quad -100{,}368{,}975.998.
\end{aligned}$$

Because we know that the discriminant is an integer, we can be relatively confident that the actual discriminant is $-100{,}368{,}976$.

Now that we worked out a difficult example; you do an easy one.

EXERCISE: Using the quadratic formula to compute the two roots of $x^2 + bx + c$, show that the discriminant of the quadratic equation, using the product formula above, is indeed $b^2 - 4c$.

SOLUTION: The quadratic formula (where $a = 1$) says that the two roots of this polynomial are $\frac{-b+\sqrt{b^2-4c}}{2}$ and $\frac{-b-\sqrt{b^2-4c}}{2}$. The formula for the discriminant says to start by computing all differences of all pairs of roots. In this case, there are only two roots, and their difference is $\sqrt{b^2 - 4c}$. The next step is to square the difference, which gives $b^2 - 4c$. The final step would be to multiply all of these squared differences together, but because there is only one number, there is no more work to be done. The discriminant is $b^2 - 4c$, as promised.

One wonderful fact is that there are formulas that give the discriminant in terms of the coefficients of the polynomial, just as in the case of the quadratic equation. Thus, although there is no general formula for the roots in terms of the coefficients, the discriminant can be computed without knowing the roots of the polynomial, and indeed the discriminant can give some information about the roots.

EXAMPLE: The discriminant of the cubic polynomial $x^3 + ax^2 + bx + c$ is the complicated expression $a^2b^2 + 18abc - 4b^3 - 27c^2 - 4a^3c$. If this expression is negative, then the cubic polynomial has exactly one real root; if positive, then the cubic polynomial has three real roots.

There is yet another new term that we need to introduce before we return to Frobenius:

DEFINITION: The *norm* of an algebraic integer is the absolute value of the constant term of its minimal polynomial. If θ is an algebraic integer, we write $N(\theta)$ for the norm of θ. In the special case of 0, we define $N(0) = 0$.

For example, if θ is any root of $x^5 + 41x^4 + 32x + 11 = 0$, then $N(\theta) = 11$.

It is a fact that $N(\alpha)N(\beta) = N(\alpha\beta)$. The definition of norm just given does not make this obvious, but it can be proved.

A Working Definition of Frob_p

Now that we have told you about the discriminant and the norm, we can tell you a little bit more about our mysterious element Frob_p. Fix a prime p. To tell you what Frob_p does to an element of $\mathbf{Q}^{\mathrm{alg}}$, it is good enough to tell you what Frob_p does to every element of $\overline{\mathbf{Z}}$, because every element of $\mathbf{Q}^{\mathrm{alg}}$ can be written as a quotient of elements of $\overline{\mathbf{Z}}$. In other words, if α is an element of $\mathbf{Q}^{\mathrm{alg}}$, then we can write $\alpha = \beta/\gamma$, where β and γ are in $\overline{\mathbf{Z}}$. So $\mathrm{Frob}_p(\alpha) = \mathrm{Frob}_p(\beta)/\mathrm{Frob}_p(\gamma)$, and all we need to do is tell you about $\mathrm{Frob}_p(\theta)$ where θ equals β or γ.

So, now suppose that θ is an algebraic integer. We have yet one more complication: We can only define $\mathrm{Frob}_p(\theta)$ if p is *unramified* with respect to θ. What does this mean? Take the minimal polynomial f of θ. Compute the discriminant Δ_f of this polynomial. If p is not a factor of Δ_f, then p is unramified with respect to θ. We can only easily define $\mathrm{Frob}_p(\theta)$ if p does not divide Δ_f.

Occasionally it can happen that p is unramified with respect to θ even if p is a factor of Δ_f. In fact, whether or not p is unramified with respect to θ is really a property of the field $\mathbf{Q}(f)$, but it is too complicated to give the accurate definition except in an appendix to this chapter. The simple criterion that p is not a factor of Δ_f will be good enough for us. Because Δ_f is just a garden variety integer, it has only finitely many prime factors. All of the other primes will be unramified with respect to θ. (Of course, which primes those will be depends on θ.) For example, if β solves the fourth-degree polynomial above, then any prime not dividing 100,368,976 will be unramified with respect to β.

Suppose that θ is a root of the $\mathbf{Z}$-polynomial $f(x)$ with leading coefficient 1. We know that $\mathrm{Frob}_p(\theta)$ has to be one of the numbers that solves the equation $f(x) = 0$. Which ones can it be?

FACT: $\mathrm{Frob}_p(\theta)$ solves the minimal polynomial for θ, and has the lovely property that the ordinary integer $N(\mathrm{Frob}_p(\theta) - \theta^p)$ is evenly divisible by p. That is, it leaves no remainder when divided by p. In particular, we are telling you that there is at least one number—call it β—that both solves the minimal polynomial for θ *and* makes $N(\beta - \theta^p)$ evenly divisible by p. In case there is only one such β, we define $\mathrm{Frob}_p(\theta)$ to be this number β. (It is possible that sometimes β will equal θ itself.) The definition of $\mathrm{Frob}_p(\theta)$ when there are several β's to choose from is given in the second appendix to this chapter.

An Example of Computing Frobenius Elements

An example, using the easiest possible nonrational element of $\overline{\mathbf{Z}}$, is $\mathrm{Frob}_p(i)$. We cannot define $\mathrm{Frob}_2(i)$, because 2 is ramified with respect to i: The minimal **Z**-equation that i satisfies is $x^2 + 1 = 0$. We know from the formula for the discriminant of a quadratic polynomial that this has discriminant -4. Therefore, Δ_f is evenly divisible by 2, and in fact 2 is ramified with respect to i, so we cannot define $\mathrm{Frob}_2(i)$.

What about $\mathrm{Frob}_3(i)$? We need to start with the lowest-degree equation that i solves, which is $x^2 + 1 = 0$. The other solution of that equation is $-i$. So $\mathrm{Frob}_3(i)$—indeed $\mathrm{Frob}_p(i)$ for any odd prime p—has to be either i or $-i$. We also need to have

$$3 | N(\mathrm{Frob}_3(i) - i^3).$$

Here we are using the handy notation $p|a$ to mean "a is evenly divisible by p." Because $i^3 = -i$, $N(\mathrm{Frob}_3(i) + i)$ has to be a multiple of 3. If we tried $\mathrm{Frob}_3(i) = i$, we would get $N(2i)$, which is not a multiple of 3. (Why not? The minimal **Z**-polynomial for $2i$ is $x^2 + 4$. It has constant term 4. Therefore, $N(2i) = 4$, and 4 is not evenly divisible by 3.) So we need to have $\mathrm{Frob}_3(i) = -i$. (*Check*: $N(-i + i) = N(0) = 0$, which is evenly divisible by 3.)

How about $\mathrm{Frob}_5(i)$? Now we get

$$5 | N(\mathrm{Frob}_5(i) - i^5)$$

which says that $5|N(\text{Frob}_5(i) - i)$. If we try $\text{Frob}_5(i) = -i$, we get $5|N(-2i)$, which is not true, so the only possibility is that $\text{Frob}_5(i) = i$.

EXERCISE: Let p be an odd prime. Show that

$$\text{Frob}_p(i) = \begin{cases} i & \text{if } p \equiv 1 \pmod 4 \\ -i & \text{if } p \equiv 3 \pmod 4). \end{cases}$$

Notice that this formula looks an awful lot like equation (7.4) on page 79. This is not a coincidence. See chapter 19.

Unfortunately, this example is misleading, for in the general case there can be more than one root of the minimal polynomial that satisfies the divisibility relationship. For the complete definition of $\text{Frob}_p(\theta)$, see the second appendix to this chapter.

We have been discussing ramification with respect to θ, and at the beginning of this chapter we spoke of ramification with respect to a Galois representation r. What is the connection? Well, we would have to spell out the connection between number fields and Galois representations, which we do in the first appendix to this chapter.

Frob_p and Factoring Polynomials modulo p

Now that we have told you what Frob_p looks like, at least some of the time, we should tell you something about it. A lot of the amazing things will wait for future chapters, but here is one thing that we can use to finish off this chapter.

Pick an irreducible **Z**-polynomial $f(x)$ of degree n with leading coefficient 1. (Irreducible just means that it does not factor into **Z**-polynomials of smaller degree. A consequence of irreducibility is that $f(x)$ is the minimal polynomial of each of its roots.) We can take the n roots $\alpha_1, \alpha_2, \dots, \alpha_n$ of this polynomial. Then we know that any element of the Galois group G permutes these roots. In particular, pick any prime p so that Δ_f is not evenly divisible by p. Then p is unramified with respect to any of the roots of $f(x)$.

We know that Frob_p permutes the roots. What can we say about this permutation?

As we have seen, permutations can be broken up into *cycles*. We can start with a root α_1, apply the permutation Frob_p to it to get another root, apply Frob_p to that root, and keep going. Eventually, we have to get back to α_1, because there are only finitely many roots. The number of different roots that we visit on our trip is called the length of a cycle. For example, if $\mathrm{Frob}_3(\alpha_1) = \alpha_2$, and $\mathrm{Frob}_3(\alpha_2) = \alpha_{11}$ and $\mathrm{Frob}_3(\alpha_{11}) = \alpha_8$, and $\mathrm{Frob}_3(\alpha_8) = \alpha_1$, then this cycle has length 4, because it visits the four different roots α_1, α_2, α_{11}, and α_8.

There will also be another cycle, starting at α_3, because α_3 was not part of the previous cycle. We can keep going, putting each root of the polynomial into a cycle, and counting the lengths of the cycles. The lengths have to add up to n, because each root is in exactly one cycle. (If you are worried about the possibility that $\mathrm{Frob}_3(\alpha_5) = \alpha_5$, we call that a cycle of length 1.) So we have a bunch of positive integers $n_1, n_2, \ldots, n_k$, so that $n_1 + n_2 + \cdots + n_k = n$. These integers are the lengths of the cycles produced by the permutation on the roots. We say that Frob_p "has cycle type $n_1 + n_2 + \cdots + n_k$."[4]

We can also try factoring the polynomial $f(x)$ in $\mathbf{F}_p$. For example, $x^2 - 11 = (x + 10)(x - 10)$ in $\mathbf{F}_{89}$. Here is an amazing theorem connecting cycle types and factorizations:

> **THEOREM 16.1**: If $f(x)$ is an irreducible **Z**-polynomial and p is a prime not dividing Δ_f, and if $f(x)$ factors in $\mathbf{F}_p$ into k factors, and the degrees of those k factors are $n_1, n_2, \ldots, n_k$, then the cycle type of Frob_p is $n_1 + n_2 + \cdots + n_k$.

In general, it is easy to factor polynomials with elements in $\mathbf{F}_p$ (there are fast computer algorithms for this), but it is difficult to figure out Frob_p. So we usually use the factorization to tell us about the cycle type of Frob_p.

---

[4]You may object that we are ignoring the ambiguity in the notation Frob_p here. However, the cycle type will turn out to be the *same*, whatever choice of Frob_p you make. Also, remember that the "+" sign in the cycle type does not denote addition, but is the traditional symbol for separating the lengths of the cycles from each other.

For instance, consider $f(x) = x^2 - 11$. The discriminant of f is 44. So consider some prime p different from 2 or 11. We use Theorem 16.1 to compute $\mathrm{Frob}_p(\sqrt{11})$. (Remember that $\sqrt{11}$ is by definition the positive square root, which is approximately 3.3166247904.)

We know Frob_p on the set of roots $\{\sqrt{11}, -\sqrt{11}\}$ is either the identity permutation (the neutral element in $\Sigma_{\{\sqrt{11},-\sqrt{11}\}}$) or the permutation that switches the two roots. In the first case, Frob_p has two cycles each of length 1, and in the second case, it has only one cycle of length 2.

Now $f(x)$ factors in $\mathbf{F}_p$ into two simpler factors if and only if 11 is a square modulo p.[5] So, by Theorem 16.1, the cycle type of Frob_p is $1+1$ if 11 is a square modulo p, and is 2 otherwise. In the first case, $\mathrm{Frob}_p(\sqrt{11}) = \sqrt{11}$ and in the second $\mathrm{Frob}_p(\sqrt{11}) = -\sqrt{11}$. For example, $\mathrm{Frob}_{89}(\sqrt{11}) = \sqrt{11}$ because $x^2 - 11$ factors into two linear factors in $\mathbf{F}_{89}$.

APPENDIX
The Official Definition of the Bad Primes for a Galois Representation

Suppose we have a Galois representation r. Remember that this means r is a morphism of $G_{\mathbf{Q}}$, the absolute Galois group of $\mathbf{Q}$ (which we are abbreviating G) to some other group H, either a group of permutations or a group of matrices.

Any morphism f from a group G to a group H has what is called a *kernel*. This is just the set of elements x in G with the property that $f(x)$ is the neutral element e of H. We call the kernel of r by the name K_r. Because finding the kernel is the same as solving the equation $f(x) = e$, you can imagine that this is an important set to study. For example, r is a faithful representation precisely when K_r contains only the neutral element e.

So K_r is a subset of G. It actually is a group all by itself, but that is not so important right now.

---

[5]That is, if and only if $11 = a^2$ for some a in $\mathbf{F}_p$.

Next we find all the algebraic numbers α with the property that they are not moved by any element in K_r. That is, let E_r be the set of all α in $\mathbf{Q}^{\text{alg}}$ such that $\sigma(\alpha) = \alpha$ for every σ in K_r. It turns out E_r is actually a field all by itself, but that is not so important right now either. So given any Galois representation r, we get a set of algebraic numbers E_r, which is called the *fixed field of the kernel of* r. Here is the payoff: The prime p is unramified with respect to r if and only if it is unramified for every α contained in E_r. The definition of this last phrase is given in the next appendix.

APPENDIX
The Official Definition of "Unramified" and Frob_p

In general, if $K = \mathbf{Q}(f)$ for some $\mathbf{Z}$-polynomial f, we now define what it means for p to be "unramified in K." If f is the minimal polynomial for θ, we also say p is unramified for θ under the same circumstances. Then we describe the set of images of Frobenius elements at p under the restriction morphism from G to $G(f)$. We just present the definitions without explanation. As we said, full explanation would require turning this appendix into a textbook on algebraic number theory.

• **DEFINITION**: The set of algebraic integers in K, written as $\mathcal{O}_K$, is defined to be $K \cap \overline{\mathbf{Z}}$.

• **DEFINITION**: An ideal of $\mathcal{O}_K$ is a subset A of $\mathcal{O}_K$ closed under addition, subtraction, and multiplication, containing 0 and satisfying the property that xa is in A whenever x is in $\mathcal{O}_K$ and a is in A.

• **DEFINITION**: If A and B are two ideals, the *product ideal* AB is the set of all sums $a_1b_1 + \cdots + a_mb_m$, where $a_1, \ldots, a_m$ are in A and $b_1, \ldots, b_m$ are in B. If k is a positive integer, A^k means the product of A with itself k times.

• **DEFINITION**: An ideal A is called a *prime ideal* if the following property holds: If x and y are in $\mathcal{O}_K$ and xy is in A, then either x or y or both must have already been in A.

• **DEFINITION**: If d is any element of $\mathcal{O}_K$, (d) is the set of all products xd where x is in $\mathcal{O}_K$. Note that (d) is an ideal.

• **THEOREM**: For any prime integer p in $\mathbf{Z}$, there are prime ideals $P_1, \ldots, P_t$ in $\mathcal{O}_K$ and a positive integer e such that $(p) = P_1^e P_2^e \cdots P_t^e$.

• **DEFINITION**: The prime number p is *unramified* in K if and only if $e = 1$.

• **THEOREM**: Let p and its ideal factorization be given as above. Suppose that p is unramified in K. Let j be an integer from 1 to t. Then there exists exactly one element σ in the Galois group $G(f)$ with the following two properties:

1. For any x in P_j, $\sigma(x)$ is again in P_j.
2. For any y in $\mathcal{O}_K$, $\sigma(y) - y^p$ is in P_j.

• **THEOREM**: The element σ is one possibility for $r_{G(f)}(\mathrm{Frob}_p)$ in $G(f)$. You get all the possibilities by varying j.

• **THEOREM**: For any β in $G(f)$, there are infinitely many primes q such that β is a possible $r_{G(f)}(\mathrm{Frob}_q)$. Thus, from the point of view of the *finite* Galois group $G(f)$, being a Frobenius element is not unusual—quite the opposite. What is difficult is knowing which q's go with which conjugacy classes in $G(f)$. That is one thing reciprocity laws are meant to help us with.

PART THREE

Reciprocity Laws

Chapter 17

RECIPROCITY LAWS

Road Map

Starting in this chapter we usually use lowercase Greek letters for Galois representations.

Suppose that we have a linear representation ϕ of the absolute Galois group G. We will produce a list of numbers by taking the trace of $\phi(\mathrm{Frob}_2)$, $\phi(\mathrm{Frob}_3), \ldots$, one number for each prime that is unramified for the representation ϕ. This list encodes a lot of crucial information about G and ϕ which can be used to study $\mathbf{Z}$-equations. The information is summarized in what are called generalized reciprocity laws.

In this chapter, we explain the general concept of a reciprocity law. In the rest of the book, we will give examples of reciprocity laws and how they work to provide information about solutions to $\mathbf{Z}$-equations. In chapter 19, we will reinterpret quadratic reciprocity as a reciprocity law in this new, generalized, sense. The connection with quadratic reciprocity is the reason we call theorems of this type *reciprocity* laws.

The List of Traces of Frobenius

A little review: We have the set of all algebraic numbers, denoted $\mathbf{Q}^{\mathrm{alg}}$. We have the absolute Galois group G of all arithmetic-preserving permutations of $\mathbf{Q}^{\mathrm{alg}}$. If R is some number system, $n \geq 1$

is some integer and $\phi : G \to \mathrm{GL}(n, R)$ is a morphism of groups, then we say that ϕ is a linear Galois representation with coefficients in R.

We have told you that each ϕ comes with a set of "bad primes," the "ramified primes." In all of the examples we will consider, this set of bad primes is finite. If p is not in this finite list of bad primes, we say that ϕ is *unramified* at p.

If ϕ is such a linear Galois representation, and σ is any element in G, then we have defined the character of ϕ at σ to be the value of the trace of $\phi(\sigma)$. Recall what this means: $\phi(\sigma)$ is an n-by-n matrix with entries in R, and its trace is the sum of the entries down the main diagonal, going from the upper-left to the lower-right. In symbols, we write $\chi_\phi(\sigma)$ for the value of this character at σ. It is a number in R.

We also have defined certain special elements of G, the Frobenius elements. They are not really single elements, but rather certain sets of elements of G, one for each prime number p. One key property they share is the following: If ϕ is unramified at p, then $\chi_\phi(\mathrm{Frob}_p)$ is a well-defined element of R. This means that it does not matter which σ you choose from the whole set of possibilities for Frob_p; $\chi_\phi(\sigma)$ is always the same number, and we call the common value $\chi_\phi(\mathrm{Frob}_p)$.

In the next few chapters we hope to show by example that this list of numbers $\chi_\phi(\mathrm{Frob}_2)$, $\chi_\phi(\mathrm{Frob}_3)$, $\chi_\phi(\mathrm{Frob}_5)$, $\ldots$ (omitting from our list those p where ϕ is ramified, in which case $\chi_\phi(\mathrm{Frob}_p)$ is not really defined), is always a very interesting list of numbers. The whole idea of reciprocity laws is to try to find other independent ways of generating these lists of numbers. If we succeed, we obtain a type of theorem called a generalized reciprocity law, or simply a reciprocity law for short.

We will be able to prove for you that a particular reciprocity law is true in only a few cases. Usually, these proofs are extremely difficult and require much additional mathematics that lies outside of the scope of this book. For the most general reciprocity laws, proofs have not yet been discovered, although the experts conjecture that the laws are true.

Black Boxes

Whatever the independent way of generating the $\chi_\phi(\mathrm{Frob}_p)$'s might be, we can think of it as a black box.[1] It is usually some complicated object from some other part of mathematics. In this book we view some of these mathematical machines that produce the sequence of numbers $\chi_\phi(\mathrm{Frob}_p)$'s as a given, the way a child can read the dial of a wristwatch and tell the time without knowing how the watch works. In some cases (e.g., quadratic reciprocity), we will be able to describe completely what is happening.

From the point of view we take in this book, a reciprocity law is a black box. You put in a prime p (where ϕ is unramified) and out pops a number. If it is the black box with the label ϕ on it, then that number will be $\chi_\phi(\mathrm{Frob}_p)$. The possession of this block box then gives us power over the list of $\chi_\phi(\mathrm{Frob}_p)$'s and over whatever it is in number theory that they tell us about, for example, torsion points on some elliptic curve, or indirectly about a supposed solution to Fermat's equation $x^n + y^n = z^n$.

You may suppose that there should be exactly one black box for each ϕ, and it should be labeled ϕ. Then if you want information about the sequence of numbers $\chi_\phi(\mathrm{Frob}_p)$, you just pull out the box labeled ϕ and start throwing in the p's.

Things are not so simple. There can be many black boxes that all give the same sequence, so they all belong to ϕ. They might be of different types (e.g., some modular forms and some cohomology classes—see chapters 20 and 21 for these concepts). Also, the black boxes do not usually come labeled by the ϕ's. Usually they are labeled with information that is derivable from some properties of ϕ. This works best if there is only a finite number of black boxes in our inventory with any particular label.

To summarize: The equality between the traces of the matrices in a Galois representation and numbers produced by some sort of black box is what is called a reciprocity law. There are various black

[1]A black box is an input–output box that you cannot see into. It may have a label on the outside, and you know what it does to the input, but you do not have to understand *why* it works. This use of the term has nothing to do with the "black box" that records flight data on an airplane.

boxes that can be used, which generally have something to do with geometry or topology. The use of these black boxes (which may be more or less transparent to the number theorist who is using them) is at the heart of the method of Galois representations in modern number theory.

Weak and Strong Reciprocity Laws

In a reciprocity law, we first specify the kind of black box we are going to consider. For instance, we specify elliptic curves. Then, for each kind of black box, there are two kinds of reciprocity law: *weak* and *strong*. In a weak reciprocity law, all we are told is that for every Galois representation ϕ of a certain type, there is some black box that outputs the corresponding $\chi_\phi(\text{Frob}_p)$'s. In the strong reciprocity law, we are also told the label on the black box. If we have not proved a reciprocity law yet, but are only guessing it is true, then we call these two different types of statements the "weak conjecture" and the "strong conjecture."

You can see that it is much easier to verify or disprove a strong conjecture, because knowledge of the label cuts down to a finite amount the number of black boxes for which you have to look. If true, the strong reciprocity law gives more information than the corresponding weak one.

It is obviously better to try to prove the strong conjecture if possible. In applications such as Wiles's proof of Fermat's Last Theorem, it was necessary to prove a strong reciprocity law to finish the proof. In terms of our analogy, the proof went like this: Suppose you have two nonzero nth powers that add up to another nth power. From this equation, you can deduce the existence of a certain Galois representation. The strong reciprocity law you have already proved implies the existence of a black box with a certain label. You work out the label of the black box. You go to your inventory of those black boxes—and there are not any with that label! Contradiction.

In summary, a strong reciprocity law tells us that the list of numbers coming from a particular Galois representation, produced by means of the Frobenius elements, will match the output from one of a certain set of black boxes (those with the correct label).

Sometimes, reciprocity laws are theorems that have been proven. Sometimes they are only conjectured and have not been proven yet. And, of course, you can mistakenly conjecture a false reciprocity law and find a numerical disproof. But the reciprocity laws in the remainder of this book are all either theorems or conjectures that most experts would say are probably going to turn out to become theorems.

Digression: Conjecture

The word conjecture means "guess." The way it is used in mathematics is "educated guess." In this digression, we will mention several conjectures from parts of mathematics other than Galois theory.

There are at least two different classes of great conjectures:

1. Those based on evidence.
2. Those based on analogy.

Some conjectures that are currently the object of intense study by mathematicians include the Poincaré Conjecture (which reportedly may have been proven true as of this writing), the Riemann Hypothesis (which is a conjecture, though it is called a "hypothesis"), and the conjecture that "$P = NP$." As a matter of fact, there is a million-dollar prize (offered by the Clay Mathematics Institute) for the first proof of any of these three conjectures, as well as some others.[2] Mathematicians may not increase their efforts because of the prize, but public interest has certainly increased.

Sometimes, serious conjectures are proven false. An example is the Mertens Conjecture. First define the Möbius function $\mu(n)$ for positive integers n: $\mu(n) = 0$ if n is divisible by a square number larger than 1; otherwise, $\mu(n) = 1$ if n has an even number of prime factors and $\mu(n) = -1$ if n has an odd number of prime factors. Then

[2]A discussion of the Millennium problems may be found in (Devlin, 2002).

you define the Mertens function for $x > 1$ by

$$M(x) = \mu(1) + \mu(2) + \mu(3) + \cdots + \mu(n),$$

where n is the largest integer less than or equal to x. The Mertens Conjecture states that for all $x > 1$,

$$|M(x)| < \sqrt{x}.$$

There is a known way to prove the Riemann Hypothesis from this statement, if it were true. If you work out $M(x)$ for small values of x, the conjecture seems to be true. In 1985, however, Andrew Odlyzko and Herman te Riele proved that the Mertens Conjecture is false.

Fermat's Last Theorem states: If x, y, and z are all nonzero integers, and n is an integer greater than 2, then it cannot happen that $x^n + y^n = z^n$. Before this theorem was proved by Andrew Wiles, it was a conjecture that fell in both classes, which is why it was regarded as particularly strong. In the simplest sense, there was evidence supporting it, as no one ever found integers solving $x^n + y^n = z^n$. The supporting evidence in fact was much stronger: For many particular values of n, mathematicians, starting in the eighteenth century, had proved that there were no solutions of $x^n + y^n = z^n$.

There was also at least one analogy that helped convince many mathematicians that Fermat's Last Theorem was true. There are many statements about integers that can also be made about polynomials, and in general the true statements about integers tend also to be true of polynomials. Proving things about polynomials is usually much simpler than proving things about integers, for at least two reasons: Polynomials have roots, and they can be differentiated. Using these tools, it is not all that difficult to show that there are no nonconstant polynomials with complex coefficients $f(x)$, $g(x)$, and $h(x)$ so that $f(x)^n + g(x)^n = h(x)^n$ if $n > 2$.

Another conjecture that is strongly supported by evidence in both categories is called the Riemann Hypothesis. This is the statement that if the real part of the complex number s is positive, then the Riemann ζ-function $\zeta(s)$ is zero only when $s = \frac{1}{2} + it$ (where t is a real number). On the one hand, you can do numerical computations and the Riemann Hypothesis looks true. On the other hand, there is a polynomial analogue to the ζ-function, and the analogous statement, called the Riemann Hypothesis for function fields, was conjectured and proved in the twentieth century.[3]

There is yet a third property of conjectures worth mentioning: They should be interesting, although two mathematicians will not always agree about what is interesting. The most interesting conjectures should imply known results, and also should imply some new and surprising results.

Kinds of Black Boxes

In the reciprocity laws or conjectures of the type we consider in this book, we always assert the identity of two sequences of numbers. One of them is always the sequence of traces of the Frobenius matrix for the unramified primes of some Galois representation. The other is a sequence derived as the output of some black box, which is usually an algebraic, geometric, or topological object. Some of these black boxes are:

1. The number of solutions to certain systems of **Z**-equations modulo various primes.
2. The Fourier coefficients of a modular form corresponding to the primes.
3. Traces of Hecke operators at the various primes.

We cover the first two cases in this book: (1) in chapters 18 and 19, (2) in chapter 21. The third case is crucial for much of the current research on generalized reciprocity laws, but unfortunately is beyond the scope of this book. For an expository article on Hecke operators in cohomology and their use in reciprocity laws, see our paper (Ash and Gross, 2000).

[3] An excellent book about the Riemann Hypothesis is (Derbyshire, 2003).

Chapter 18

ONE- AND TWO-DIMENSIONAL REPRESENTATIONS

Road Map

We are nearing our journey's end. From here forward, we will give examples of Galois representations, reciprocity laws, and their applications to Diophantine equations. In this chapter, we discuss representations of the absolute Galois group G to $\mathrm{GL}(1, \mathbf{F}_p)$ and $\mathrm{GL}(2, \mathbf{F}_p)$. We explain the first case in detail. The second example depends on elliptic curves, and we must content ourselves with a brief description and a lengthy example. The example gives us an opportunity to put together some of the facts we have learned about elliptic curves, factoring polynomials modulo q, and Frob_q. We will see how they interact to enable us to *prove* a reciprocity law concerning the 2-torsion points on an elliptic curve.

Roots of Unity

We start with the polynomial $x^n - 1$. We know what the roots of this polynomial are, using *de Moivre's Theorem*: They are the n complex numbers ζ, ζ^2, ζ^3, $\ldots$, ζ^n, where $\zeta = \cos\left(\frac{2\pi}{n}\right) + i \sin\left(\frac{2\pi}{n}\right)$. It is important to know that these n numbers are all unequal, that they all solve the same polynomial equation $x^n - 1 = 0$, and that

$\zeta^n = 1$. One consequence of these facts is that $a \equiv b \pmod{n}$ if and only if $\zeta^a = \zeta^b$.[1]

We look now at what the elements of the absolute Galois group G can do to these numbers. A key point is that when we know what σ, an element of G, does to ζ, we know what it does to every root of the polynomial $x^n - 1$. For example, suppose that $\sigma(\zeta) = \zeta^3$. Then $\sigma(\zeta^2) = \sigma(\zeta \cdot \zeta) = \sigma(\zeta) \cdot \sigma(\zeta) = \zeta^3 \cdot \zeta^3 = \zeta^6 = \zeta^{3 \cdot 2}$, and similarly $\sigma(\zeta^k) = \zeta^{3k}$ for any positive integer k.

So if we know $\sigma(\zeta)$, then we know everything that we need to know about how σ permutes the roots. Because $\sigma(1) = 1$, $\sigma(\zeta)$ must be a root not equal to 1 (i.e., $\zeta, \zeta^2, \ldots, \zeta^{n-2}$ or ζ^{n-1}). We can write $\sigma(\zeta) = \zeta^a$ for some a between 1 and $n - 1$. To distinguish elements of G, we label them by the exponent a, so that $\sigma_a(\zeta) = \zeta^a$. Of course, there will be many σ's in G with the same effect on ζ, and we will write σ_a for any one of them that takes ζ to ζ^a.

Which numbers a can occur? Suppose there is some number $d > 1$ so that both a and n are multiples of d. In that case, σ_a is expelled from the G-club, because it will not be a permutation of the roots. For example, let $n = 10$, so ζ is a "primitive" tenth root of unity.[2] We claim that σ_2 is not a permutation. What goes wrong? $\sigma_2(\zeta) = \zeta^2$, and $\sigma_2(\zeta^6) = \zeta^{12} = \zeta^{10+2} = \zeta^{10}\zeta^2 = 1 \cdot \zeta^2 = \zeta^2$. Because σ_2 sends both ζ and ζ^6 to the same number ζ^2, it follows that σ_2 is not a one-to-one correspondence. In general:

> **THEOREM 18.1**: If a and n have a common factor other than 1, then there is no σ_a in G. But if a and n have no common factor other than 1, then there are σ_a's in G.

Now we will take the case where $n = p$, a prime number bigger than 2. (This restriction to odd primes is reasonable, because $x^2 - 1$ has only the boring rational roots 1 and -1.) We can now think of the exponent a as a nonzero element of $\mathbf{F}_p$, in other words as an

---

[1]To review what $a \equiv b$ means, look back at chapter 4.

[2]The n roots of $x^n - 1$ form a cyclic group under multiplication, and to say that ζ is "primitive" means that ζ generates the entire group. See page 40 for a review of cyclic groups.

element of $\mathbf{F}_p^\times$.[3] Remember that now our defining equation for ζ is $\zeta^p = 1$.

The Galois group $G(x^p - 1)$ has exactly $p - 1$ elements: $\sigma_1, \ldots, \sigma_{p-1}$. On the other hand, any element of $\mathbf{Q}(x^p - 1)$ is of the form $c_1\zeta + c_2\zeta^2 + \cdots + c_{p-1}\zeta^{p-1}$ for some rational numbers $c_1, \ldots, c_{p-1}$. If you want, you can completely work out how G acts on the field $\mathbf{Q}(x^p - 1)$.

EXERCISE: Given a number a not divisible by p, show that there is a permutation π of $\{1, 2, \ldots, p - 1\}$ such that

$$\begin{aligned}\sigma_a(c_1\zeta + c_2\zeta^2 + \cdots + c_{p-1}\zeta^{p-1}) &= c_1\zeta^a + c_2\zeta^{2a} + \cdots + c_{p-1}\zeta^{(p-1)a} \\ &= c_{\pi(1)}\zeta + c_{\pi(2)}\zeta^2 + \cdots + c_{\pi(p-1)}\zeta^{p-1}\end{aligned}$$

so that the c_i's get permuted by this permutation π (where π depends on a).

How Frob$_q$ Acts on Roots of Unity

We next pick some prime q other than p, and we try to identify Frob_q as one of these permutations of the roots of $x^p - 1$.

THEOREM 18.2: If $q \neq p$, then q is unramified in $\mathbf{Q}(x^p - 1)$, and Frob_q is a σ_q. In other words, if ζ is a primitive pth root of unity, then $\text{Frob}_q(\zeta) = \zeta^q$.

We can explain why this theorem is plausible, even though we cannot quite prove it here. We start with a numerical example of the easiest case: $p = 3$. The solutions of the equation $x^3 - 1 = 0$ are $x = 1, x = \frac{-1+\sqrt{-3}}{2}$, and $x = \frac{-1-\sqrt{-3}}{2}$. One way to get these solutions is to apply de Moivre's Theorem; another is to factor $x^3 - 1 = (x - 1)(x^2 + x + 1)$ and apply the quadratic formula to solve $x^2 + x + 1 = 0$. To avoid writing these numbers over and over, it is customary to write $\omega = \frac{-1+\sqrt{-3}}{2}$. You can check that $\omega^2 = \frac{-1-\sqrt{-3}}{2}$, so the three solutions to the equation are 1, ω, and ω^2. Remember that $\omega^3 = 1$.[4]

---

[3]See page 39 for a discussion of $\mathbf{F}_p^\times$.

[4]Thus, when $p = 3$, we write ω rather than ζ.

Let q be any prime other than 3, and let us see what $\mathrm{Frob}_q(\omega)$ is. The two key facts to remember are that $\mathrm{Frob}_q(\omega)$ must be a root of the same minimal polynomial as ω, which is x^2+x+1, and that q is a divisor of $N(\mathrm{Frob}_q(\omega)-\omega^q)$. The two roots of x^2+x+1 are ω and ω^2, so we must have $\mathrm{Frob}_q(\omega)=\omega$ or $\mathrm{Frob}_q(\omega)=\omega^2$. It is also easy to see that if $\mathrm{Frob}_q(\omega)=\omega^q$, then $N(\mathrm{Frob}_q(\omega)-\omega^q)=N(0)=0$, and q does divide 0. So setting $\mathrm{Frob}_q(\omega)=\omega^q$ does not contradict either one of the requirements for $\mathrm{Frob}_q(\omega)$ that we stated back on page 185. (Remember that ω^q must be ω or ω^2, because $\omega^3=1$; therefore, powers of ω just keep repeating over and over again.)

The other possibility is that $\omega^q=\omega^j$, where j is 1 or 2, and $\mathrm{Frob}_q(\omega)=\omega^k$, where k is $3-j$. (This is a fancy way of writing that k is also 1 or 2, and $k\neq j$.) Because $\omega^3=1$, we can compute that with these assumptions, $N(\mathrm{Frob}_q(\omega)-\omega^q)=N(\omega^{3-j}-\omega^j)$, where j is 1 or 2. If $j=1$, you get $N(\omega^2-\omega)=N(-\sqrt{-3})=3$, and if $j=2$, you get $N(\omega-\omega^2)=N(\sqrt{3})=3$. Either way, there is no way for the prime q to divide $N(\omega^{3-j}-\omega^j)$. So we need to have

$$\mathrm{Frob}_q(\omega)=\omega^q. \tag{18.3}$$

This same argument works if 3 is replaced by any odd prime p. The last step is a bit trickier in general: You need to show that if $j+k$ is not a multiple of p, then $N(\zeta^{p-k}-\zeta^j)$ is p, which can be done with some algebraic identities.

Now we explore one particular case of Theorem 18.2. We look at what happens when $q\equiv 1 \pmod p$. This means that $\mathrm{Frob}_q(\zeta)=\zeta$. Why? We can rewrite the congruence $q\equiv 1 \pmod p$ as the equation $q=1+kp$, so $\zeta^q=\zeta^{1+kp}=(\zeta^1)(\zeta^{kp})=\zeta\cdot(\zeta^p)^k=\zeta\cdot(1)^k=\zeta$. In other words, Frob_q will be the identity permutation. Now, remember what we said in chapter 16 about thinking of Frob_q as a permutation of the roots of an irreducible polynomial. When $q\equiv 1 \pmod p$, Frob_q is a permutation that does not change any of the roots, so each of the cycles in the permutation has size 1. And, by Theorem 16.1, this means that x^p-1 must factor into p factors of degree 1 in $\mathbf{F}_q$.

For example, let $p=5$ and $q=11$. Then by what we wrote in the previous paragraph, because $11\equiv 1 \pmod 5$, x^5-1 must factor into five factors in $\mathbf{F}_{11}$. In fact, you can check that

$$x^5 - 1 \equiv (x-1)(x-3)(x-4)(x-5)(x-9) \pmod{11}. \tag{18.4}$$

The congruence (18.4) means that if you multiply out the five factors on the right and then view each coefficient of the result modulo 11, you will get $x^5 - 1$. *Check*: The product of those five factors is $x^5 - 22x^4 + 176x^3 - 638x^2 + 1{,}023x - 540$, and 22, 176, 638, and 1,023 are all multiples of 11, while $-540 = -49 \times 11 - 1 \equiv -1 \pmod{11}$.

On the other hand, if you take $q = 7$ (for example), you *will not* be able to find five integers $b_1, \ldots, b_5$ so that

$$x^5 - 1 \equiv (x-b_1)(x-b_2)(x-b_3)(x-b_4)(x-b_5) \pmod{7}.$$

One-Dimensional Galois Representations

We can now construct some one-dimensional representations of G. The number a in the equation $\sigma_a(\zeta) = \zeta^a$ can be thought of as a 1-by-1 matrix, that is, an element of $\mathrm{GL}(1, \mathbf{F}_p)$. The fact that a has a multiplicative inverse modulo p is exactly what makes a an element of $\mathrm{GL}(1, \mathbf{F}_p)$, rather than just an element of $\mathbf{F}_p$.

We describe this one-dimensional Galois representation very carefully. After all, it is the first Galois representation we have been able to understand completely in this book. We call it ϕ. If γ is any element in the absolute Galois group G, $\phi(\gamma)$ is going to be some element in $\mathrm{GL}(1, \mathbf{F}_p)$, in other words, $\phi(\gamma) \equiv a \pmod{p}$, for some integer a not divisible by p. What is $\phi(\gamma)$?

First of all, $\phi(\gamma)$ will only depend on the restriction of γ to the field $\mathbf{Q}(f)$ where $f(x) = x^p - 1$. We call the Galois group of $\mathbf{Q}(f)$, which we normally denote by $G(f)$, H for short. Then $\phi(\gamma)$ depends only on $r_H(\gamma)$. (Look back at chapter 14 for the details of the restriction morphism.)

We saw in our recent discussion that every element of H is of the form σ_a for some a not divisible by p. So $r_H(\gamma) = \sigma_a$ for some a not divisible by p. We simply define $\phi(\gamma)$ to be that a.

In other words, you can always figure out what $\phi(\gamma)$ is by applying γ to ζ: $\phi(\gamma) = a$ if and only if $\gamma(\zeta) = \zeta^a$, or, concisely,

$$\gamma(\zeta) = \zeta^{\phi(\gamma)}.$$

Using this formula, you can easily check that ϕ is a morphism. Say $\phi(\gamma) = a$ and $\phi(\gamma') = a'$. Then what is $\phi(\gamma \circ \gamma')$? Use our test, and apply $\gamma \circ \gamma'$ to ζ: $\gamma \circ \gamma'(\zeta) = \gamma(\gamma'(\zeta)) = \gamma(\zeta^{a'}) = \gamma(\zeta)^{a'}$ (since γ respects all arithmetic operations, including raising to a power) $= (\zeta^a)^{a'} = \zeta^{aa'}$ by the laws of exponents. So the test tells us that $\phi(\gamma \circ \gamma') = aa'$. But this is equal to $\phi(\gamma)\phi(\gamma')$. So $\phi(\gamma \circ \gamma') = \phi(\gamma)\phi(\gamma')$ for all elements γ and γ' in G—and that is the defining property of a morphism.

So ϕ really is a representation of G. It is a linear representation to $\mathrm{GL}(1, \mathbf{F}_p)$. Because we are using only 1-by-1 matrices, we call it a "one-dimensional" representation. So $\chi_\phi(\mathrm{Frob}_q)$ is the same as $\phi(\mathrm{Frob}_q)$.[5] Finally, a prime q is unramified in $\mathbf{Q}(x^p - 1)$ as long as $q \neq p$, and the reciprocity law that we get from Theorem 18.2 says that

$$\chi_\phi(\mathrm{Frob}_q) = \phi(\mathrm{Frob}_q) \equiv q \pmod{p}.$$

This is a brilliant example of a reciprocity law. The black box labeled ϕ simply outputs q when you input q.

Two-Dimensional Galois Representations Arising from the p-Torsion Points of an Elliptic Curve

We now switch our attention to elliptic curves. Pick an elliptic curve E, and let p be some prime. We have written earlier about the set $E[p]$, the p-torsion of the elliptic curve. These are the points P on $E(\mathbf{C})$ that solve the equation $pP = \mathcal{O}$. (This equation should be thought of as similar to the equation $x^p = 1$ above.) Remember that the set $E[p]$ has a structure: We can find two particular elements P and Q in this set so that every element in $E[p]$ can be written as $aP + bQ$, where the numbers a and b are between 0 and $p - 1$ (and

[5] Remember that the trace of the 1-by-1 matrix $[q]$ is just the number q itself.

so we can think of them as element of $\mathbf{F}_p$). If σ is any element of G, we can associate a matrix to σ as on page 147; we will now call that matrix $\psi(\sigma)$.

We claim that ψ is a representation of G, this time to $\mathrm{GL}(2, \mathbf{F}_p)$. If σ is any element in the absolute Galois group G, $\psi(\sigma)$ depends only on the restriction of σ to the field $\mathbf{Q}(E[p])$ generated by the coordinates of the p-torsion points of the elliptic curve E. So you can see what 2-by-2 matrix ψ sends σ to by applying σ to P and to Q and writing down the answers in terms of P and Q. If $\sigma(P) = aP + bQ$ and $\sigma(Q) = cP + dQ$, then $\psi(\sigma)$ is the matrix $\begin{bmatrix} a & c \\ b & d \end{bmatrix}$.

Again, you can check that ψ is a morphism. It is a little more complicated, involving multiplication of two 2-by-2 matrices. When you try to check this, remember that applying $\sigma \circ \sigma'$ to P and Q means first applying σ' and then applying σ.

So ψ is a linear representation of G to $\mathrm{GL}(2, \mathbf{F}_p)$. It is a two-dimensional linear Galois representation.

In the case of the one-dimensional linear representation ϕ above, we saw that $\phi(\mathrm{Frob}_q)$ tells us something very interesting about the prime q, namely, it tells us how the polynomial $x^p - 1$ factors over $\mathbf{F}_q$, which further tells us things such as whether $\mathbf{F}_q$ contains a primitive pth-root of unity. Similarly, ψ tells us some very interesting things about the elliptic curve E. For example, if $h(x)$ is the $\mathbf{Z}$-polynomial whose roots are the first coordinates of all the p-torsion points on E, then $\psi(\mathrm{Frob}_q)$ gives information about how $h(x)$ factors modulo q.

But there is much more to know about an elliptic curve. The Galois representation ψ, and similar (but more complicated) constructions that use *all* the torsion points of $E(\mathbf{Q}^{\mathrm{alg}})$, can tell us about that extra information. For example, they can tell us whether E has extra symmetries of its own, called "complex multiplications." Also, they can be used to make predictions about whether there are infinitely many rational points in $E(\mathbf{Q})$, summed up in what is known as the Birch–Swinnerton-Dyer Conjecture, named after the two number theorists Bryan Birch and Peter Swinnerton-Dyer. These things are unfortunately too advanced to explain in this book, but it is good to remember that no matter how difficult or deep is something we do discuss, there are plenty of more difficult and much deeper facts awaiting discussion. For a little more about this, see chapters 21 and 22.

How Frob_q Acts on p-Torsion Points

Note that because we made some choices, namely, of P and Q, the matrix $\psi(\sigma)$ could change if our choices changed, but it will stay in the same conjugacy class. Hence, as we know from chapter 16, the trace of this matrix will not change. In other words, the character $\chi_\psi(\sigma)$ is an element of $\mathbf{F}_p$ that can be computed and will stay the same, regardless of how we choose P and Q.

Here now is an amazing theorem connecting $\chi_\psi(\text{Frob}_q)$ and $E(\mathbf{F}_q)$, the set of solutions modulo q of the equation $y^2 = x^3 + Ax + B$ that defines the elliptic curve E (plus the point $\mathcal{O}$ as usual):

> **THEOREM 18.5**: Let q be a prime other than p that is unramified for the Galois representation ψ. (This will be true if q does not divide $2p(4A^3 + 27B^2)$.) The matrix $\psi(\text{Frob}_q)$ is only defined up to conjugacy, but $\chi_\psi(\text{Frob}_q)$ is well-defined, and
>
> $$\chi_\psi(\text{Frob}_q) \equiv 1 + q - \#E(\mathbf{F}_q) \pmod{p}.$$
>
> Here $\#E(\mathbf{F}_q)$ means the number of points in $E(\mathbf{F}_q)$. In other words, $\#E(\mathbf{F}_q)$ equals the number of solutions to the congruence $y^2 \equiv x^3 + Ax + B \pmod{q}$ defining the elliptic curve E, plus 1 (for $\mathcal{O}$, the point at infinity).

Now we are keeping track of the rather subtle information of whether or not $x^3 + Ax + B$ is a square modulo q! The formula above is both strikingly interesting and only the tip of an iceberg. There are similar formulas, although much more complicated, for all $\mathbf{Z}$-varieties. Consider what this formula means in the case of an elliptic curve. Given the elliptic curve, we get this two-dimensional Galois representation ψ. It has to do with the p-torsion on the elliptic curve. From ψ, we take only its trace, χ_ψ. When we apply it to Frob_q we get a number closely related to the number of mod q solutions to the cubic equation that defines the elliptic curve!

We use the elliptic curve E and the prime p to compute a machine (the Galois representation), put Frob_q through the machine, and get out an interesting fact about q. We could just as well start out with any other p (as long as it is not equal to q) and get the "same" fact about q, because after all $\#E(\mathbf{F}_q)$ depends only on q, not on p. (We put "same" in quotes because we are looking at the same number $1+q-\#E(\mathbf{F}_q)$ but modulo different primes p.) It's funny: The Galois representation ψ does depend on p. But when we apply χ_ψ to Frob_q we get the "same answer": $1+q-\#E(\mathbf{F}_q) \pmod p$, no matter what p is![6]

If we let p vary and for each p take the corresponding ψ, we get what is called a compatible family of Galois representations, because they have the "same" traces on Frobenius elements. This is a general fact about étale cohomology of **Z**-varieties (see chapter 19).

You can read a formula such as $\chi_\psi(\text{Frob}_q) \equiv 1+q-\#E(\mathbf{F}_q) \pmod p$ in one of two ways. It can tell you what $\#E(\mathbf{F}_q) \pmod p$ is, if you know enough about Frob_q, or if you know $\#E(\mathbf{F}_q) \pmod p$, it will tell you something about Frob_q. In practice, it is used in both ways.

We can also view this formula as a reciprocity law, where the black box is E, used to produce the number $1+q-\#\mathbf{F}(E_q)$, which in turn tells us the trace of Frob_q acting via the Galois representation ψ. In fact, more knowledge about elliptic curves sometimes enables us to find the actual conjugacy class of $\psi(\text{Frob}_q)$ after we know the number $1+q-\#\mathbf{F}(E_q)$.

As we said already, similar comments apply to viewing the formula $\chi_\phi(\text{Frob}_q) \equiv q \pmod p$ as a reciprocity law for the simple one-dimensional Galois representation ϕ. Here the black box was much simpler: Drop in q and get out q modulo p.

---

[6]This does not contradict the fact that an irreducible representation of a group is determined by its characteristic polynomial, because that fact applies only to two representations over the *same* field. Here, as we vary p, the field $\mathbf{F}_p$ varies. It just happens that there is one "divine" integer $1+q-\#E(\mathbf{F}_q)$ whose "avatars" (reductions mod p) are showing up.

The 2-Torsion

We conclude this chapter with a relatively simple case where we can actually prove Theorem 18.5, using various facts stated earlier in the book. Namely, we will look at the Galois representation ψ coming from an elliptic curve where we set $p = 2$. We will be able to compute $\psi(\mathrm{Frob}_q)$ explicitly (which is much more difficult if $p > 2$) and test the theorem: For q any prime except for divisors of $2(4A^3 + 27B^2)$, we must have the number of points on $E(\mathbf{F}_q)$ odd or even depending on whether $\chi_\psi(\mathrm{Frob}_q)$ is odd or even.

Pick an elliptic curve E defined by the equation $y^2 = x^3 + Ax + B$. We are letting $p = 2$ in the above theorem, meaning that we want to identify the points on E that satisfy the equation $2P = \mathcal{O}$.[7] It happens that these points are particularly easy to describe. In addition to $\mathcal{O}$ itself, they are the points $P = (a, 0)$, $Q = (b, 0)$, and $R = (c, 0)$ where a, b, and c all solve the equation $x^3 + Ax + B = 0$. (This is a cubic equation, so it will have three solutions. One of the consequences of the restriction that $4A^3 + 27B^2 \neq 0$ is that the three solutions will be unequal, so there really are three *different* points in our list.) It will always happen that $P + Q = R$.

Be forewarned that some of the analysis gets quite complicated, and you can skip it if you just want to go on to the next chapter. But if you plow through it, you should get an idea of how many of the ideas in this chapter and chapter 16 fit together beautifully.

An Example

We take a particularly easy example to get started understanding the theorem. Let E be the elliptic curve defined by the equation $y^2 = x^3 - x$ (in other words, we are taking $A = -1$ and $B = 0$). To find the points in $E[2]$, we need to solve the equation $x^3 - x = 0$. This equation has three solutions: $x = 0$, $x = 1$, and $x = -1$. Corresponding to those solutions there are three points on the elliptic curve: $P = (0, 0)$, $Q = (1, 0)$, and $R = (-1, 0)$. If you use the

[7]Remember that this means $P + P = \mathcal{O}$, where the plus sign means adding points according to the recipe given in chapter 9.

complicated group law on E, you can check that $P+P=\mathcal{O}$, $Q+Q=\mathcal{O}$, and $R+R=\mathcal{O}$. In fact, you can also check that $P+Q=R$.

Now, let q be any odd prime. We want to understand how Frob_q permutes these three points. In this case, we can easily understand what happens: Frob_q is an element of the absolute Galois group G, and the absolute Galois group leaves fixed the rational numbers 0, 1, and -1. In other words, $\text{Frob}_q(P)=P$, $\text{Frob}_q(Q)=Q$, and $\text{Frob}_q(R)=R$.[8]

In this case, the matrix $\psi(\text{Frob}_q)$ is well-defined, and using the prescription on page 206, we see that

$$\psi(\text{Frob}_q)=\begin{bmatrix}1 & 0\\ 0 & 1\end{bmatrix}.$$

The trace of this matrix is 2. Therefore, $\chi_\psi(\text{Frob}_q)=2$, and Theorem 18.5 is the assertion that

$$2\equiv q+1-\#E(\mathbf{F}_q) \pmod 2.$$

Because $2\equiv 0 \pmod 2$, this is a fancy way of saying that the number $q+1-\#E(\mathbf{F}_q)$ is even. Because q is an odd prime, $q+1$ is even, so in the end we have to understand why the number of points in $E(\mathbf{F}_q)$ is even.

How can we arrange the points on $E(\mathbf{F}_q)$ to make them easy to count? Sort them into three classes:

1. $\mathcal{O}$.
2. P, Q, and R.
3. The rest of the solutions of $y^2\equiv x^3-x \pmod q$.

We can see that there is one element in the first class and three elements in the second class. What about the third class? The other solutions of $y^2\equiv x^3-x \pmod q$ will have $y\not\equiv 0 \pmod q$. If a typical point in this class is (j,k), then $(j,-k)$ will also be in this class and will be different from (j,k), because $k\not\equiv 0 \pmod q$.

Thus, because we can pair them up, there are an even number of points in the third class. There are four other points on the

[8]The third equation in fact is implied by the other two, and the fact that $P+Q=R$. Why? If σ is any element of the absolute Galois group G, then $\sigma(P+Q)=\sigma(P)+\sigma(Q)$ if P and Q are any points on the elliptic curve $E=(\mathbf{Q}^{\text{alg}})$.

curve—$\mathcal{O}$, P, Q, and R—so in all there must be an even number of points on the curve.

Another Example

We now try a slightly more complicated example. Let E be the elliptic curve $y^2 = x^3 - 1$, so that $A = 0$ and $B = -1$. We can apply the theorem for any prime q other than 2 and 3. The three points in $E[2]$ (other than $\mathcal{O}$) are $P = (1, 0)$, $Q = (\omega, 0)$, and $R = (\omega^2, 0)$, where ω is the number $\frac{-1+\sqrt{-3}}{2}$ that showed up above in our discussion of the solutions of $x^3 - 1 = 0$.

EXERCISE: Show that $P + Q = R$.

Now choose your favorite prime q (other than 2 and 3, which are ramified), and consider how Frob_q shifts these points around. We know that $\text{Frob}_q(P) = P$, because the coordinates of P are in $\mathbf{Z}$, and elements of the absolute Galois group G always fix elements of $\mathbf{Z}$. What about $\text{Frob}_q(Q)$? The two possibilities are $\text{Frob}_q(Q) = Q$ and $\text{Frob}_q(Q) = R$.

We take $q = 5$. We saw above in equation (18.3) that $\text{Frob}_5(\omega) = \omega^5 = \omega^2$, and so $\text{Frob}_5(Q) = R$. Because $R = P + Q$, we see, using the prescription on page 206 as usual, that

$$\psi(\text{Frob}_5) = \begin{bmatrix} 1 & 1 \\ 0 & 1 \end{bmatrix}.$$

Therefore, $\chi_\psi(\text{Frob}_5) = 1 + 1 \equiv 0 \pmod 2$.

What about Frob_7? Repeat the same computation, and you will see that $\text{Frob}_7(\omega) = \omega$. Therefore, $\text{Frob}_7(Q) = Q$, and so

$$\psi(\text{Frob}_7) = \begin{bmatrix} 1 & 0 \\ 0 & 1 \end{bmatrix}.$$

This means that $\chi_\psi(\text{Frob}_7) = 1 + 1 \equiv 0 \pmod 2$.

In fact, you can see that these two examples show the two possibilities for $\psi(\text{Frob}_q)$, and in either case $\chi_\psi(\text{Frob}_q) \equiv 0 \pmod 2$. Again, from Theorem 18.5, we conclude that $\#E(\mathbf{F}_q)$ is always even.

How can we see that there are always an even number of points in $\#E(\mathbf{F}_q)$? The argument is very similar to the one that we used for the elliptic curve $y^2 = x^3 - x$. In this case, we can divide up the points on E into three types:

1. $\mathcal{O}$.
2. Elements of $E(\mathbf{F}_q)$ of the form $(x, 0)$.
3. The rest of $E(\mathbf{F}_q)$.

We know that there is exactly one point in the first set. The element $(1, 0)$ is always in the second set, so the question is whether there is one element in the second set or three (depending on whether or not -3 is a perfect square in $\mathbf{F}_q$). Either way, there are an even number of elements between the first two sets, and the third set always has an even number of elements, for exactly the same reason as before: If (j, k) is a point on the elliptic curve, so is $(j, -k)$.

Yet Another Example

We now consider a more complicated example. Let E be the elliptic curve $y^2 = x^3 - 2$. In this case, the three solutions to $x^3 - 2 = 0$ are $\sqrt[3]{2}$, $\omega\sqrt[3]{2}$, and $\omega^2\sqrt[3]{2}$. The elements of $E[2]$ (other than $\mathcal{O}$) are $P = (\sqrt[3]{2}, 0)$, $Q = (\omega\sqrt[3]{2}, 0)$, and $R = (\omega^2\sqrt[3]{2}, 0)$. Again, you can check that $P + Q = R$.

Pick a prime q. We would like to know how Frob_q shifts these three points around, which is the same as knowing how Frob_q shifts around the three roots of $x^3 - 2$. For any particular q, we can work this out, but we appeal instead to Theorem 16.1. That is the theorem that tells us that we can find out the cycle type of Frob_q as a permutation of the roots of $x^3 - 2$ by factoring $x^3 - 2$ modulo q. There are three possibilities:

(1) $x^3 - 2$ factors into three linear factors modulo q. For example, if $q = 31$, you can check that

$$x^3 - 2 \equiv (x + 11)(x + 24)(x + 27) \pmod{31}$$

by multiplying out $(x+11)(x+24)(x+27) = x^3 + 62x^2 + 1209x + 7128$ and reducing that polynomial modulo 31. In this case, the cycle type of Frob_q is $1+1+1$.

We can also tell from this factorization that $(-11)^3 \equiv 2 \pmod{31}$ and $(-24)^3 \equiv 2 \pmod{31}$ and $(-27)^3 \equiv 2 \pmod{31}$. In other words, we know that there are three points on $E(\mathbf{F}_{31})$ whose second coordinate is 0: $(-11,0)$, $(-24,0)$, and $(-27,0)$.

(2) $x^3 - 2$ factors into a linear and a quadratic factor modulo q. For example, if $q = 11$, then

$$x^3 - 2 \equiv (x+4)(x^2+7x+5) \pmod{11},$$

where $x^2 + 7x + 5$ does not factor any more modulo 11. You can check the factorization by computing $(x+4)(x^2+7x+5) = x^3 + 11x^2 + 33x + 20$ and reducing modulo 11. In this case, the cycle type of Frob_q is $1+2$.

We can also tell from this factorization that $(-4)^3 \equiv 2 \pmod{11}$, and that there is one point on $E(\mathbf{F}_{11})$ that has a second coordinate of 0: $(-4,0)$.

(3) $x^3 - 2$ does not factor at all modulo q. For example, if $q = 7$, $x^3 - 2$ does not factor modulo 7. In this case, the cycle type of Frob_q is 3.

This factorization tells us that in $E(\mathbf{F}_7)$, there are *no* points with a second coordinate of 0.

A cycle type of $1+1+1$ means that Frob_q does not shift around the roots of the polynomial. So we know that Frob_{31} fixes the roots of $x^3 - 2$, which in turn means that $\mathrm{Frob}_{31}(P) = P$, $\mathrm{Frob}_{31}(Q) = Q$, and $\mathrm{Frob}_{31}(R) = R$. Therefore,

$$\psi(\mathrm{Frob}_{31}) = \begin{bmatrix} 1 & 0 \\ 0 & 1 \end{bmatrix}$$

and $\chi_\psi(\mathrm{Frob}_{31}) = 1 + 1 \equiv 0 \pmod 2$.

We can also tell that there are an even number of points in $E(\mathbf{F}_{31})$. There is $\mathcal{O}$, and there are three points with a second coordinate of 0, and there must be an even number of points that have a nonzero second coordinate.

A cycle type of $1+2$ means that Frob_q fixes one of the three roots of the polynomial, and shifts around the other two. So we know that

Frob_{11} fixes one of the three points P, Q, and R, and shifts the other two. We cannot know which one is fixed, but we can rename them so that P is fixed and $\text{Frob}_{11}(Q) = R$. Then

$$\psi(\text{Frob}_{11}) = \begin{bmatrix} 1 & 1 \\ 0 & 1 \end{bmatrix}$$

and $\chi_\psi(\text{Frob}_{11}) = 1 + 1 \equiv 0 \pmod 2$.

Again, we can tell that there are an even number of points on $E(\mathbf{F}_{11})$. There is $\mathcal{O}$, and there is one point with a second coordinate of 0, and there must be an even number of points that have a nonzero second coordinate.

Finally, a cycle type of three means that Frob_q shifts all three of the roots of the polynomial, and therefore all three of the points. We do not know if $\text{Frob}_7(P) = Q$ or $\text{Frob}_7(P) = R$, but we can again rename them so that $\text{Frob}_7(P) = Q$ and $\text{Frob}_7(Q) = R$. Then

$$\psi(\text{Frob}_7) = \begin{bmatrix} 0 & 1 \\ 1 & 1 \end{bmatrix}$$

and $\chi_\psi(\text{Frob}_7) = 0 + 1 \equiv 1 \pmod 2$.

Now we can tell that there are an *odd* number of points on $E(\mathbf{F}_7)$. There is $\mathcal{O}$, and there are no points with a second coordinate of 0, and, as always, there are an even number of points with a nonzero second coordinate.

The Proof

In general, how can we verify that $\chi_\psi(\text{Frob}_q) \equiv 1 + q - \#E(\mathbf{F}_q) \pmod 2$ for any elliptic curve $y^2 = x^3 + Ax + B$? First, notice that $1 + q$ is always even, so we have to check that $\chi_\psi(\text{Frob}_q) \equiv \#E(\mathbf{F}_q) \pmod 2$.

Again, we arrange the points on $E(\mathbf{F}_q)$ into three classes:

1. $\mathcal{O}$.
2. Other elements of $E(\mathbf{F}_q)$ that are in $E[2]$, that is, elements $P = (j, k)$ in $E(\mathbf{F}_q)$ with $k = 0$.
3. The remainder of the points in $E(\mathbf{F}_q)$.

As before, we know that there are always an even number of points in the third category, and there is always one point in the first category.

What can we say about the second category? These points correspond to solutions of $x^3 + Ax + B \equiv 0 \pmod{q}$, which in turn correspond to cycles of length 1 in the factorization above. For example, on the elliptic curve $y^2 = x^3 - 2$ when $q = 31$, there are three points in the second category, corresponding to the cycle structure $1+1+1$. When $q = 11$, there is one point in the second category corresponding to the cycle structure $1+2$. And when $q = 7$, there are no points in the second category corresponding to the cycle structure 3.

To summarize: A cycle structure of types $1+1+1$ or $1+2$ corresponds to an *even* number of points on $E(\mathbf{F}_q)$, and a cycle structure of type 3 corresponds to an *odd* number of points.

Back to the representation: A cycle structure of type $1+1+1$ corresponds to the matrix representation

$$\begin{bmatrix} 1 & 0 \\ 0 & 1 \end{bmatrix},$$

a cycle structure of type $1+2$ corresponds to the matrix representation

$$\begin{bmatrix} 1 & 1 \\ 0 & 1 \end{bmatrix},$$

and a cycle structure of type 3 corresponds to the matrix representation

$$\begin{bmatrix} 0 & 1 \\ 1 & 1 \end{bmatrix}.$$

Therefore, the cycle structure that predicts even and odd number of points exactly corresponds to $\chi_\psi(\mathrm{Frob}_q)$.

Chapter 19

QUADRATIC RECIPROCITY REVISITED

Road Map

In this chapter we perform a lot of algebra to reinterpret quadratic reciprocity. As explained in chapter 7, quadratic reciprocity appears to be an innocuous curiosity, the main part of which states that if p and q are two odd primes, then the "squareness" of p modulo q and the "squareness" of q modulo p are related in a certain predictable way. Throughout the nineteenth century, number theorists proved similar theorems about "cubedness" and higher powers. In the beginning of the twentieth century, mathematicians began to realize that all of these power reciprocity theorems could be clarified and unified in terms of Galois representations. The key was to think of a reciprocity law as a way of describing $\chi_\phi(\mathrm{Frob}_p)$ by other means, as we explained in chapter 17.

Here we undertake this interpretation for quadratic reciprocity—the simplest, the oldest, and the most fundamental of all reciprocity laws. We also must bring **Z**-equations into the discussion, particularly the equation $x^2 - W = 0$ for various integers W. This is because asking about the "squareness" of p modulo q is the same as asking whether $x^2 - p = 0$ has a solution modulo q.

The black boxes we use in this chapter are introduced in the first few paragraphs. As we go through the algebra, we will be able to see into these black boxes, another sign that quadratic reciprocity is relatively easy.

Simultaneous Eigenelements

Let us return to one-dimensional representations, and show how we can use the concept of a generalized reciprocity law to deduce quadratic reciprocity. The black box associated with the Galois representations we consider in this chapter is constructed from an integer $N > 1$ and a set of functions $\mathscr{F}_N$ with some special properties. First we have to make the following definition:

> **DEFINITION**: Let N be an integer greater than 1. Then $(\mathbf{Z}/N\mathbf{Z})^\times$ is the set of all integers b modulo N that have multiplicative inverses modulo N.

For example, 3 is in $(\mathbf{Z}/10\mathbf{Z})^\times$ because $3 \cdot 7 \equiv 1 \pmod{10}$, but 5 is not in $(\mathbf{Z}/10\mathbf{Z})^\times$ because there is no solution to the congruence equation $5x \equiv 1 \pmod{10}$. If b is in $(\mathbf{Z}/N\mathbf{Z})^\times$, we write b^{-1} for the element in $(\mathbf{Z}/N\mathbf{Z})^\times$ that satisfies the congruence $bb^{-1} \equiv 1 \pmod{N}$.

A theorem that is proved in most introductory number theory courses says that the elements of $(\mathbf{Z}/N\mathbf{Z})^\times$ are the numbers between 1 and $N-1$ that have no factor in common with N other than 1.

Now we can define the set $\mathscr{F}_N$:

> **DEFINITION**: Let N be an integer greater than 1. Then $\mathscr{F}_N$ stands for the set of all functions from $(\mathbf{Z}/N\mathbf{Z})^\times \to \mathbf{C}^\times$.[1]

If α is a function in $\mathscr{F}_N$, and if b is an element of $(\mathbf{Z}/N\mathbf{Z})^\times$, then as usual with functional notation, we write $\alpha(b)$ for the complex number in $\mathbf{C}^\times$ that the rule α assigns to b.

The set $\mathscr{F}_N$ contains the black box. To describe these black boxes, we need to consider certain functions inside of this set.

> **DEFINITION**: Let N be an integer greater than 1. A *simultaneous eigenelement* in $\mathscr{F}_N$ is a function α in $\mathscr{F}_N$ with

---

[1]Recall that $\mathbf{C}^\times$ denotes the set of all nonzero complex numbers.

the property that for each prime p that is not a divisor of N, there is a complex number a_p such that

$$\alpha(p^{-1}x) = a_p\alpha(x), \tag{19.1}$$

if x is any element of $(\mathbf{Z}/N\mathbf{Z})^\times$. (Note that $p^{-1}x$ is also in $(\mathbf{Z}/N\mathbf{Z})^\times$, because of the assumption that p is not a divisor of N.) The numbers $a_2, a_3, \ldots$ are called the *eigenvalues* corresponding to the function α.

An equation similar to (19.1) turns out to be true in many generalized reciprocity laws. Unfortunately, the equation is usually more complicated than this one.

Equation (19.1) has surprising consequences which we will explore in the remainder of this chapter. We can start here with an observation. Suppose that p and q are two primes that satisfy the congruence $p \equiv q \pmod{N}$. Then $p^{-1} \equiv q^{-1} \pmod{N}$, and therefore if x is any integer at all, $p^{-1}x \equiv q^{-1}x \pmod{N}$. This means that $\alpha(p^{-1}x) = \alpha(q^{-1}x)$, because α is a function from $(\mathbf{Z}/N\mathbf{Z})^\times$. The right-hand side of equation (19.1) now tells us that $a_p\alpha(x) = a_q\alpha(x)$, and, because $\alpha(x)$ is nonzero, we can conclude that $a_p = a_q$.

This is worth restating: If $p \equiv q \pmod{N}$, then $a_p = a_q$. This equality is a special property of eigenvalues that occur in the study of one-dimensional Galois representations.

The Z-Variety $x^2 - W$

What do these eigenvalues mean?

A reciprocity law would tell us that this list $a_2, a_3, \ldots$ should have something to do with the traces of Frobenius elements Frob_2, $\text{Frob}_3, \ldots$ under some Galois representation ϕ. It turns out that this is true, and that we get the correct ϕ by studying the $\mathbf{Z}$-variety $x^2 - W = 0$ for some nonzero integer W. We assume that W is not a perfect square. Then any element σ in the absolute Galois group G must take a solution of $x^2 - W = 0$ to a solution of the same equation. We designate one of the solutions as $\sqrt{W}$, and then the other is $-\sqrt{W}$. Hence, $\sigma(\sqrt{W}) = \sqrt{W}$ or $\sigma(\sqrt{W}) = -\sqrt{W}$. This allows us to define a function ϕ from G to $\{+1, -1\}$ by the rule

$$\sigma(\sqrt{W}) = \phi(\sigma)\sqrt{W}.$$

You can easily check that ϕ is a morphism from G to $\{+1, -1\}$.[2] Because $+1$ and -1 are numbers, we can view ϕ as a Galois representation from G to $\mathrm{GL}(1, \mathbf{C})$, viewing numbers as being the same as 1-by-1 matrices, as we have done before.

Because the trace of a 1-by-1 matrix is the same as the number in the matrix, we see that the character χ_ϕ of the representation ϕ is just ϕ. To emphasize this, we will use the symbol χ rather than ϕ.

To summarize: For any nonsquare integer W, we have the Galois representation χ from G to $\mathrm{GL}(1, \mathbf{C})$ defined by

$$\sigma(\sqrt{W}) = \chi(\sigma)\sqrt{W}.$$

We really should include the letter W somewhere in this notation, because χ depends on W, but the notation $\chi^{(W)}$ is too cluttered.

From the definition of Frob_p, or from the fact that the cycle type of $\chi(\mathrm{Frob}_p)$ is related to how $x^2 - W$ factors modulo p,[3] follows this next fact:

THEOREM 19.2: If p is an odd prime that does not divide W, p is unramified for χ and

$$\chi(\mathrm{Frob}_p) = \left(\frac{W}{p}\right).$$

Remember that $\left(\frac{W}{p}\right)$ is called the *Legendre symbol*. It does *not* refer to dividing W by p. The equation in the theorem is at least

[2]The group law on $\{+1, -1\}$ is multiplication.

[3]The set of elements Frob_p in the absolute Galois group of $\mathbf{Q}$ gives information about how polynomials factor in $\mathbf{F}_p$ and vice versa. If $\left(\frac{W}{p}\right) = 1$, then the congruence $y^2 \equiv W \pmod{p}$ has two solutions, say a and b. Then the polynomial $x^2 - W$ factors as $x^2 - W \equiv (x-a)(x-b) \pmod{p}$. On the other hand, if $\left(\frac{W}{p}\right) = -1$, then the congruence $y^2 \equiv W \pmod{p}$ has no solutions, and the polynomial $x^2 - W$ does not factor into the product of two linear polynomials modulo p. Now, by Theorem 16.1, in the first case, the cycle type of $\chi(\mathrm{Frob}_p)$ is $1+1$ so that $\chi(\mathrm{Frob}_p)$ must be the neutral element 1 in the multiplicative group $\{+1, -1\}$. In the second case, the cycle type of $\chi(\mathrm{Frob}_p)$ is 2 so that $\chi(\mathrm{Frob}_p)$ must be the other element -1 in the group.

believable, because if you look back at the definition of $\chi(\sigma)$, you will see that $\chi(\sigma)$ has to be 1 or -1.

A Weak Reciprocity Law

THEOREM 19.3 (*A Weak Reciprocity Law*): Given a square-free integer W,[4] there is some positive integer N and a simultaneous eigenelement α in $\mathcal{F}_N$ with eigenvalues $a_2, a_3, \ldots$ such that

$$\chi(\mathrm{Frob}_p) = a_p$$

for every prime p that is not a divisor of N. (If p is not a divisor of N, then p is unramified for the Galois representation χ.)

Why is this a reciprocity law? It tells us how to find the list of numbers $\chi(\mathrm{Frob}_p)$ in terms of the black box α. This black box is labeled N. It does not tell us exactly what α is (although we will be able to figure that out later), but the assertion of the mere existence of α gets us going. We will need to strengthen this law a bit, and then it will imply quadratic reciprocity.

To see how this reciprocity law implies quadratic reciprocity, we first have to discover what the eigenvalues are for a simultaneous eigenelement α in $\mathcal{F}_N$. So let us suppose that α is a simultaneous eigenelement. We may multiply α by any constant in $\mathbf{C}^\times$ and it will still satisfy equation (19.1), with the same eigenvalues. If we multiply α by a constant c which has the property that $c\alpha(1) = 1$, we get a function that sends 1 to 1. In other words, henceforth we can assume that $\alpha(1) = 1$.

From our definitions, we know that $\alpha(p^{-1}b) = a_p\alpha(b)$ for any number b in $(\mathbf{Z}/N\mathbf{Z})^\times$. Set $b = p$, and we have $\alpha(p) = a_p^{-1}$. Next, take the formula $\alpha(p^{-1}b) = a_p\alpha(b)$, and let $b = py$. We get $\alpha(y) = a_p\alpha(py)$. Move a_p to the other side of the equation, and we have

---

[4]A *square-free* integer is one that is not divisible by any perfect square other than 1. For example, 30 is square-free, and 12 is not square-free.

$\alpha(py) = a_p^{-1}\alpha(y)$. We know that $a_p^{-1} = \alpha(p)$, so make that one final substitution, and we get the equation $\alpha(py) = \alpha(p)\alpha(y)$. If x is any number in $(\mathbf{Z}/N\mathbf{Z})^\times$, we can apply this formula one by one to all of the prime factors of x, and conclude that $\alpha(xy) = \alpha(x)\alpha(y)$.

In other words, the eigenelement α is a *morphism* of $(\mathbf{Z}/N\mathbf{Z})^\times$ to $\mathbf{C}^\times$, that is,

$$\alpha(xy) = \alpha(x)\alpha(y).$$

Moreover, if α is an eigenelement from Theorem 19.3 (for some W), $\alpha(p) = a_p^{-1} = \chi(\text{Frob}_p)^{-1} = \chi(\text{Frob}_p)$ because $\chi(\text{Frob}_p) = 1$ or -1. Note also that $a_p^{-1} = a_p$ for the same reason.

Summarize all of this as the following:

THEOREM 19.4: If α is a simultaneous eigenelement of $\mathscr{F}_N$ satisfying Theorem 19.3 such that $\alpha(1) = 1$, then α is a morphism from $(\mathbf{Z}/N\mathbf{Z})^\times$ to $\mathbf{C}^\times$, and the eigenvalues are given by the formula

$$a_p = \alpha(p) = \chi(\text{Frob}_p) = \left(\frac{W}{p}\right).$$

The last two equalities follow from Theorems 19.2 and 19.3 above. In particular we have the formula

$$\alpha(p) = \left(\frac{W}{p}\right)$$

which we will use freely from now on.

A Strong Reciprocity Law

To be able to derive quadratic reciprocity, we need to strengthen our Weak Reciprocity Law. Given W, we have to tell you what N to use.

THEOREM 19.5 (*Stronger Version of the Reciprocity Law*): In Theorem 19.3, we can let $N = 4\,|W|$.

Here, $|W|$ means as usual the absolute value of W (e.g., $|-4| = 4$ and $|7| = 7$). If you are patient enough, you can untangle what this theorem and all of the preceding definitions are describing. You will find out that it is quite close to Theorem 7.2, which we used to prove quadratic reciprocity. Theorem 19.5 gives us a different approach. To prove quadratic reciprocity starting from Theorem 19.5, we take clever choices of W, and use the existence of the eigenelement α that Theorem 19.3 gives us, together with the knowledge of N given by Theorem 19.5.

In this section, we will be varying W, so we will put a subscript on the corresponding α given to us by Theorems 19.3 and 19.5, writing it as α_W. That is, α_W is the simultaneous eigenelement depending on W described in Theorem 19.4. Because $N = 4|W|$, we have the important fact that if $a \equiv b \pmod{4|W|}$, then $\alpha_W(a) = \alpha_W(b)$. We will use this fact several times in our derivation of quadratic reciprocity.

A Derivation of Quadratic Reciprocity

Suppose first that $W = -1$, so that $N = 4$. Because $\alpha_W(p) = \left(\frac{-1}{p}\right)$ can be thought of as a morphism defined on $(\mathbf{Z}/4\mathbf{Z})^\times$, we can conclude that $\left(\frac{-1}{p}\right)$ is determined just by the value of $p \pmod 4$. Because $\left(\frac{-1}{3}\right) = -1$, and $\left(\frac{-1}{5}\right) = 1$, we can deduce the usual formula for $\left(\frac{-1}{p}\right)$ given on page 79. For example, if $p \equiv 3 \pmod 4$, $\left(\frac{-1}{p}\right) = \left(\frac{-1}{3}\right) = -1$. See chapter 7, if you have not already glanced back to refresh your memory.

Next, take $W = 2$, and we know that α is a morphism defined on $(\mathbf{Z}/8\mathbf{Z})^\times$. Computation of $\left(\frac{2}{p}\right)$ for $p = 3, 5, 7$, and 17 gives the usual formula for $\left(\frac{2}{p}\right)$, which we gave on page 79.

Next, suppose that p and q are the odd primes that we want to compare in quadratic reciprocity. First, assume that $p \equiv q \pmod 4$, with $p > q$, and let $W = (p - q)/4$. Then $p \equiv q \pmod{4W}$, which means (because of Theorem 19.5) that $\alpha_W(p) = \alpha_W(q)$, or $\left(\frac{W}{p}\right) = \left(\frac{W}{q}\right)$. We have

$$\begin{aligned}\left(\frac{p}{q}\right) &= \left(\frac{4W+q}{q}\right) = \left(\frac{4W}{q}\right)\\ &= \left(\frac{W}{q}\right) = \left(\frac{W}{p}\right) = \left(\frac{4W}{p}\right) = \left(\frac{p-q}{p}\right)\\ &= \left(\frac{-q}{p}\right) = \left(\frac{-1}{p}\right)\left(\frac{q}{p}\right),\end{aligned}$$

which implies quadratic reciprocity for $p \equiv q \equiv 1 \pmod 4$ and $p \equiv q \equiv 3 \pmod 4$.

To derive the remaining cases, observe that for any W the congruence

$$x^2 - W \equiv 0 \pmod{4W - 1}$$

always has the solution $x \equiv 2W$. This tells us that if W is positive and p is any prime dividing $4W - 1$, then $\alpha_W(p) = 1$. Factoring $4W - 1$ into a product of primes, and using the fact that α_W is a morphism, we see that $\alpha_W(4W - 1) = 1$. In other words, we have $\alpha_W(-1) = 1$.

Now, suppose that $p + q \equiv 0 \pmod 4$, and let $W = (p + q)/4$. Then $\alpha_W(p) = \alpha_W(4W - q) = \alpha_W(-q) = \alpha_W(-1)\alpha_W(q) = \alpha_W(q)$, which implies that $\left(\frac{W}{p}\right) = \left(\frac{W}{q}\right)$, and, reasoning as before, we can conclude that $\left(\frac{p}{q}\right) = \left(\frac{q}{p}\right)$.

Note that Theorem 19.4 actually identified the mysterious simultaneous eigenelement α. It is just the Legendre symbol:

$$\alpha_W(p) = \left(\frac{W}{p}\right).$$

In the case of more general reciprocity laws, we usually do not understand the black box this well.

The fact that the Galois representation χ was also closely related to the Legendre symbol is a sort of coincidence occurring because quadratic reciprocity is the simplest of all reciprocity laws. Something similar happens with other one-dimensional Galois representations of G. This relationship is called class field theory. But as soon as you go to higher-dimensional Galois representations, the black boxes diverge radically from the Galois representations in their definitions and become more complicated too. This starts to happen in the study of representations arising from elliptic curves.

> **EXAMPLE**: Let $W = -1$ and p be an odd prime. Then Frob_p either fixes or exchanges the roots $i, -i$ of $x^2 + 1$. The fact that $\chi(\mathrm{Frob}_p) = \left(\frac{-1}{p}\right)$ means that $\mathrm{Frob}_p(i) = \left(\frac{-1}{p}\right) i$. This explains the connection between equation (7.4) in chapter 7 and the last exercise in chapter 16.

Chapter 20

A MACHINE FOR MAKING GALOIS REPRESENTATIONS

Road Map

We now jump from specific examples of Galois representations to an important general method for finding them. This method, using "étale cohomology," sets up a direct relationship between systems of **Z**-equations and representations of the absolute Galois group G. This relationship comes from some very advanced mathematical constructions whose nature we can describe only in a very sketchy way. The discovery (or invention) of étale cohomology is one of the major successes of twentieth-century mathematics.

Vector Spaces and Linear Actions of Groups

We have seen in preceding chapters examples of one- and two-dimensional Galois representations. But it is very unlikely that they include all the information to be found in the absolute Galois group of **Q**. Indeed they do not. We seek a large supply of Galois representations of all dimensions.

We need a general method to derive linear Galois representations from **Z**-varieties. There are several of these methods, but we discuss just one: *étale cohomology*. It is too advanced to explain in detail here, but we can give a very rough idea of what is going on. First, a definition:

DEFINITION: If k is a field and $n \geq 1$, then an *n-dimensional vector over k* is an n-by-1 matrix with entries from k. The set of all n-dimensional vectors over k is denoted k^n, and is called a *vector space*.

For example, $\mathbf{C}^3$ contains the three-dimensional vector

$$\begin{bmatrix} 3 \\ i \\ \sqrt{2} \end{bmatrix}.$$

The vector

$$\begin{bmatrix} 2 \\ 4 \\ 3 \\ -10 \end{bmatrix}$$

is contained in $\mathbf{Q}^4$, and is also in $\mathbf{R}^4$.

DEFINITION: A *line* in k^n is a subset of k^n consisting of all vectors that are proportional to some given vector.

An example of a line in $\mathbf{C}^3$ is the set of all the three-dimensional vectors

$$\begin{bmatrix} 3c \\ ic \\ \sqrt{2}c \end{bmatrix},$$

where c takes on all possible values in $\mathbf{C}$ (including 0—the 0-vector is in every line by this definition.)

A linear representation, you will remember, is a morphism from a group G to a group of square matrices $\mathrm{GL}(n, k)$, where k is some number system. We call this a "linear" representation because if M is a matrix and L is a line in the n-dimensional space of vectors k^n, then ML is always a line again.

For example, if M is the matrix $\begin{bmatrix} 1 & 2 \\ 4 & 5 \end{bmatrix}$ in $\mathrm{GL}(2, \mathbf{Q})$ and L is the line of all vectors $\begin{bmatrix} a \\ b \end{bmatrix}$ such that $b = 2a$ (if you graphed this, it would be the line of slope 2 through the origin) then ML is the line of all vectors $\begin{bmatrix} c \\ d \end{bmatrix}$, where $d = (14/5)c$.

Why? When we multiply the matrix M by the vector $\begin{bmatrix} a \\ b \end{bmatrix}$ we get $\begin{bmatrix} a+2b \\ 4a+5b \end{bmatrix}$. Now if $\begin{bmatrix} a \\ b \end{bmatrix}$ were a point on the line L to begin with, that would mean that $b = 2a$. If we substitute this in, we get $\begin{bmatrix} a+2b \\ 4a+5b \end{bmatrix} = \begin{bmatrix} a+4a \\ 4a+10a \end{bmatrix} = \begin{bmatrix} 5a \\ 14a \end{bmatrix}$. This last vector is of the form $\begin{bmatrix} c \\ d \end{bmatrix}$ where $d = \frac{14}{5}c$.

Thus M sends the line of slope 2 to the line of slope $\frac{14}{5}$. Similarly, you can check that M sends any line through the origin to another line through the origin. And this is why we call $v \to Mv$ a *linear map*.

Now suppose M is any n-by-n matrix in $\mathrm{GL}(n, k)$ and v and w are any vectors in k^n, and a is any number in k. Then $M(v + w) = Mv + Mw$ and $M(av) = aMv$. These formulas follow from the distributive and associative laws of matrix multiplication.

Conversely, if you have any permutation s of k^n that always satisfies the formulas $s(v + w) = s(v) + s(w)$ and $s(av) = as(v)$, then there is some matrix M so that s is the same as multiplication by M: $s(v) = Mv$ for every v in k^n. Of course, M depends on s.

So one way of constructing linear representations of a group G is by finding a vector space k^n and a *linear action* of G on it. This is by definition a morphism f from G to the permutation group of k^n that satisfies the formulas $s(v + w) = s(v) + s(w)$ and $s(av) = as(v)$ for every permutation $s = f(g)$ coming from any group element g in G.

For example, think about the group A_4 of symmetries of a tetrahedron which we discussed in chapter 12. There we defined a morphism r with the property that for any element σ in A_4, $r(\sigma)$ is the rotation of space that causes the permutation σ on the vertices of the tetrahedron. You can think of a rotation of space as a permutation of $\mathbf{R}^3$. As such, it is a linear action on $\mathbf{R}^3$ because its effect as a permutation of $\mathbf{R}^3$ can be given as multiplication by a certain matrix in $\mathrm{GL}(3, \mathbf{R})$. (The particular matrix depends, of course, on the particular rotation.) This gives a linear action of A_4 on $\mathbf{R}^3$: For any σ in A_4, $r(\sigma)$ acts on $\mathbf{R}^3$ by rotating it.

Linearization

We have seen that G, the absolute Galois group of $\mathbf{Q}$, has a permutation representation on the roots of any given $\mathbf{Z}$-polynomial. Similarly, G has a permutation representation on the $\mathbf{Q}^{\mathrm{alg}}$-points of any $\mathbf{Z}$-variety W, defined by permuting all of the solutions in $\mathbf{Q}^{\mathrm{alg}}$ of the system of equations defining W. This works fine when studying $\mathbf{Z}$-varieties defined by a single polynomial in a single variable. But when the varieties get more complicated, these permutation representations are very hard to work with. We want to relate them to some *linear* representations of G to get a better handle on the number-theoretic properties of the variety W.

What is needed is a way to replace the permutation representation of G on the $\mathbf{Q}^{\mathrm{alg}}$ points of W by a related linear representation of G on some set of vectors associated with W. This is what étale cohomology does. The process of replacing a complicated object by a simpler linear object is called *linearization*. It is what must be done to $\mathbf{Z}$-varieties in order to get the Galois representations we want.

For a much simpler example of this process of linearization, consider the old problem of figuring out your momentary speed while driving along a highway. If you are driving at a constant speed c, and you graph your distance driven versus the time, you will get a straight line with slope c. So the slope is your speed. But if you are speeding up and slowing down, the graph will be a curve. What is your speed exactly *now* (say at $t = 10.2$ seconds after you started)?

To answer this question you have to replace the curved graph by the straight line that approximates it most closely just at time $t = 10.2$. This is the tangent line to the graph at that point. It has a slope, and that slope is your instantaneous speed at time 10.2.

We say that we have linearized the graph of distance versus time at that point by replacing it with the tangent line. If you are interested only in your motion in a very small time frame around $t = 10.2$, then this line will give a good approximation. This idea can be souped up to study all kinds of motion governed by differential equations, as studied in calculus.

The idea of linearization has also been used by topologists. Consider a doughnut sitting in space, or more precisely the surface of the doughnut. Suppose it is a very smooth surface, say an old-fashioned without frosting. This is a somewhat wavy surface and hard to study exactly, because the equations that describe it are very complicated. Topologists defined the *tangent plane* at each point of the surface. Each tangent plane is nice and linear, and some problems about the surface can be translated into questions of linear representations of a certain group (a *monodromy group*) on these tangent planes. Then topologists went even further and figured out how to linearize some of the topological structure of surfaces, for example, how many holes they have. The vector spaces they invented are called "cohomology groups." A doughnut has only one hole, but a pretzel can have more than one.

Although it is not difficult to count the holes in a real pretzel in your hand, prior to eating it, when a surface pops out of an abstract mathematical construction it can be very difficult to figure out its properties, such as how many holes it has. The cohomology groups can help us to do so.

Étale Cohomology

Algebraic geometers were able to make an amazing translation of these topological ideas to varieties, coming up with étale cohomology. Let W be a $\mathbf{Z}$-variety. Choose a prime number p. There is a field called $\mathbf{Q}_p$, too complicated to be defined here.[1] It is a sort of cross between $\mathbf{F}_p$ and $\mathbf{Q}$. For each non-negative integer j, there is a set of vectors defined over $\mathbf{Q}_p$ denoted $H^j(W, \mathbf{Q}_p)$, which is called the jth étale cohomology group of W. It comes with a linear action of G, the absolute Galois group of $\mathbf{Q}$. It is related algebraically to W in a way

[1] If you know how to write numbers in base p, you can define $\mathbf{Q}_p$ as the set of all "infinite p-adic expansions" $\cdots a_3a_2a_1a_0.b_1b_2b_3\cdots b_r$. These are like decimal expansions of numbers in $\mathbf{R}$, except that they are allowed to be infinite *to the left but not to the right*, and are in base p instead of base 10. We leave it to you to figure out how to add and multiply them, or you can see (Koblitz, 1984).

that is similar to the way that topological cohomology groups are related to topological surfaces.

For example, if E is an elliptic curve, then $H^1(E, \mathbf{Q}_p)$ is very closely related to the mod p linear representation of $\mathbf{Q}$ studied in chapter 18. What we considered in that chapter corresponds only to the mod p version of this étale cohomology, namely, $H^1(E, \mathbf{F}_p)$, but the whole construction can be lifted up from $\mathbf{F}_p$ to $\mathbf{Q}_p$.

To give you some idea of what $\mathbf{Q}_p$ is, we can say this. Replacing $\mathbf{F}_p$ by $\mathbf{Q}_p$ turns out to be equivalent, in the case of elliptic curves, to considering all the p^m-torsion points on E, for $m = 1, 2, 3, \ldots$. If we do this, we can generalize the formula

$$\chi_\psi(\mathrm{Frob}_q) \equiv 1 + q - \#E(\mathbf{F}_q) \pmod p. \tag{20.1}$$

from chapter 18 in such a way that it determines $\#E(\mathbf{F}_q)$ exactly, not just modulo p. This is related to the fact that if there is some integer x and you know what x is modulo p^m for all m, then you can figure out what x is exactly (by taking m large enough.)

One nifty thing is that the same $\mathbf{Z}$-variety W has an entire sequence of étale cohomology groups as p runs through all the possible prime numbers. But the basic number theory, for instance, the trace of Frob_q for unramified primes q, will be "independent" of p in the following sense. The trace of Frob_q in the representation of G acting on $H^j(W, \mathbf{Q}_p)$ is a number in $\mathbf{Q}_p$. Because the fields $\mathbf{Q}_p$ are all different for different p's, it seems as if we could not compare these traces as p varies. But it turns out that the trace of Frob_q is always an *algebraic integer*—it is a root of some polynomial $f_p(x)$ with $\mathbf{Z}$-coefficients and leading coefficient 1. (We should have told you that $\mathbf{Z}$ is a subset of $\mathbf{Q}_p$ for every p.) So we can compare the traces by comparing the polynomials $f_p(x)$. If we take care to use the polynomial of smallest possible degree, we will see that it is always the same, independently of which p we are using. We call the set of Galois representations on $H^j(W, \mathbf{Q}_p)$ as p varies a "compatible family."

Here is some skippable detail about étale cohomology. Let q be a prime number, which we will assume to be unramified in the Galois representations we consider. For any integer $m \geq 1$, there is a field with exactly q^m elements, called $\mathbf{F}_{q^m}$. This is one of the first things you prove in an undergraduate course in Galois theory, and in fact these fields were first discovered by Galois. If $m = 1$, we get back our old friend $\mathbf{F}_q$.

Now let W be a $\mathbf{Z}$-variety. We can then look at the solutions of the equations defining W where we substitute elements from $\mathbf{F}_{q^m}$. In other words we look at $W(\mathbf{F}_{q^m})$. It turns out that the set of numbers $\#W(\mathbf{F}_{q^m})$ is closely connected to the matrices giving the action of Frob_q on $H^j(W, \mathbf{Q}_p)$ as we let $j = 0, 1, 2, \ldots$. This fact is a vast generalization of equation (20.1) for elliptic curves. It gives us a concrete link between the étale cohomology of W and solving the $\mathbf{Z}$-equations that define W.

You may object that this link only goes to solutions where the variables are set equal to elements of these weird finite fields, $W(\mathbf{F}_{q^m})$. What about the solutions we are really interested in, that is to say, what about $W(\mathbf{Q})$? This is one of the main current topics of research. Suppose we ask: Is $W(\mathbf{Q})$ finite or infinite? Current research suggests looking at $\#W(\mathbf{F}_{q^m})$ for *all* q and m. If this number tends to be bigger than average (in a certain precise, technical sense) then you can conjecture that it is because $W(\mathbf{Q})$ has lots of elements, and their reduction modulo q for all the different q's is what accounts for the bigger than average size of the $\#W(\mathbf{F}_{q^m})$'s. In the case of elliptic curves, this statement is called the Birch–Swinnerton-Dyer Conjecture.

End of skippable detail.

Conjectures about Étale Cohomology

The étale cohomology of $\mathbf{Z}$-varieties furnishes an abundance of linear representations of G. These representations all share certain technical properties that make it possible to work with them, although very much about them is still unknown. The conjecture of Jean-Marc Fontaine and Barry Mazur states that any linear representation of G over $\mathbf{Q}_p$ that shares these technical properties actually results as part of an étale cohomology group of some $\mathbf{Z}$-variety W. This is a powerful conjecture that is very far from having been proved. It is safe to say that no one has any idea how to prove it in general.

You can see how the Fontaine–Mazur Conjecture may have been made: You think of every way possible to do something—in this case finding linear Galois representations that satisfy the nice properties—and then you conjecture that there are no other ways. If you are smart enough, your conjecture may hold true, because you are so smart and you were not able to think of any other ways of doing it.

Another way of making conjectures is to have the intuition that some analogy should hold. This is how étale cohomology was discovered. Very roughly speaking, André Weil conjectured that some such cohomology theory must exist, based on the analogy from topology. Then Alexander Grothendieck figured out a construction, still based on the topological analogy, that gave the theory a reality. Then Pierre Deligne proved many of its most important properties. This all happened in the middle of the last century, over a period of 20 or 30 years.

At the moment, étale cohomology, and some other kinds of similar cohomology theories, are the most general known ways of constructing "nice" linear representations of the absolute Galois group G of $\mathbf{Q}$. It seems that further study of these cohomology theories, in tandem with a deepening knowledge of G, will be the most likely direction in which the study of generalized reciprocity in number theory will go. We expect there to be beautiful reciprocity laws involving the action of Frobenius elements on étale cohomology groups and modular forms (see chapter 21) and their generalizations, called "automorphic representations." (Yes—more representations!)

Chapter 21

A LAST LOOK AT RECIPROCITY

Road Map

In this chapter we discuss the idea of a "generalized reciprocity law" more broadly. We put it in the context of mathematical patterns. Looking again at the examples of general reciprocity that we described in earlier chapters, and introducing some new examples, we attempt to give the flavor of the most general reciprocity laws currently conjectured. They form part of what is commonly known as the "Langlands program."

What Is Mathematics?

A fair definition of mathematics might be "the logical study of patterns":

- Patterns of numbers—look at the chapters on quadratic reciprocity for some complicated but beautiful examples.
- Patterns of permutations—for example, the behavior of cycle decompositions.
- Patterns of points—this is geometry.
- Patterns of solutions to systems of **Z**-equations—this is what this book is all about.

We use the term "pattern" in the broadest possible sense, to mean any arrangement of things that follows some orderly rule, allowing for prediction or contemplation.

A simple pattern is the empty set.[1] If you know the solution to some problem is the empty set, you may know a lot. For example, "What is the set of all eternal truths?" We are not saying the solution is the empty set, but if it is, it might be nice to know it. Of course, this is not a mathematical problem.

"What are all possible ways of arranging your dinner guests around a circular table so no one sits next to his or her spouse?" Such patterns are the province of the field of mathematics called *combinatorics*, and are also very important in probability theory.

"What are all solutions to the **Z**-equation $x^n + y^n = z^n$, where x, y, and z are positive integers and n is an integer greater than 2?" We know now that the solution is the empty set. That is the conjecture of Fermat, now proved by Wiles.

But how did he prove it? Roughly speaking, mathematicians had already proved that if there were *any* solutions, there would be some pattern—in fact, a certain two-dimensional linear representation of the absolute Galois group G. Wiles showed that this particular pattern could not exist. In fact he proved something much more important—a certain reciprocity law connected with elliptic curves and modular forms.[2] That reciprocity law then had the corollary that the Galois representation coming from a supposed solution of $x^n + y^n = z^n$ could not exist. We will explain this in more detail in chapter 22.

So, sometimes, even to show that a certain pattern is very simple (e.g., empty), you have to know about what other sorts of patterns can exist. Two basic types of problems in mathematics can be phrased as follows:

1. Can a pattern with certain given properties exist or not?
2. Classify all patterns with the given properties.

The first statement can be called "the existence question" and the second statement the "classification problem."

The reason mathematics has so many applications to other fields, such as physics, chemistry, biology, economics, and so on, is that

[1]Perhaps only a mathematician would think of the empty set as a pattern—the "null pattern."

[2]See later in this chapter for a discussion of modular forms.

mathematicians must think in patterns. If we can describe some phenomena carefully enough as a pattern, then mathematics may be able to give information about that kind of pattern.

There is a big debate as to whether logic is part of mathematics or mathematics is part of logic. We use logic to think. We notice that our thinking, when it is valid, goes in certain patterns. These patterns can be studied mathematically. Thus, logic is a part of mathematics, called "mathematical logic." There are some amazing theorems here, such as Gödel's Theorems[3] and the theorem on algorithmic unsolvability of Diophantine equations mentioned near the end of chapter 6. On the other hand, because logic in some sense encompasses all valid thought, you may prefer to say that mathematics is a part of logic.

Another example of patterns: What is a group? It is just a pattern that certain things can exhibit when you have a composition law for always getting a third thing by combining any two others. Then the same group-pattern can show up in pure mathematics, particle physics, crystallography, and so on.

Reciprocity

A generalized reciprocity law is the bringing together of two patterns. One pattern is the set of traces, or more generally the characteristic polynomials (if you know what these are), of Frob_p acting in a Galois representation. The other pattern comes from the black box—another mathematical object of some different type. The law is like a mirror in which you can see the pattern better than in the original. But it is a two-way street. You can consider the Frob's as the original problem and the black box as the easier thing to get a handle on, or vice versa. Sometimes both sets of patterns are well understood and the reciprocity law boils down to a very beautiful symmetry of numbers—as in the case of quadratic reciprocity.

---

[3]We suggest (Hofstadter, 1979; Nagel and Newman, 2001; Smullyan, 1992) as possible starting points to learn about this.

As our last example of a black box, we consider *modular forms*. Although modular forms were discovered or invented long before étale cohomology and have a lot of different applications in mathematics and physics, they can also be used in reciprocity laws in conjunction with the étale cohomology of certain **Z**-varieties. Although the theory of étale cohomology gives us a huge realm to think about, often we cannot say much unless we get another type of gadget, much more explicit and easy to work with, that in particular situations contains the same information. Besides modular forms, other such black boxes are *finite flat group schemes* and *formal groups*.[4]

Modular Forms

Modular forms are explained in various books about Wiles's proof of Fermat's Last Theorem (Hellegouarch, 2002; Singh, 1997; van der Poorten, 1996). For our purposes, you can think of a modular form as an infinite series in powers of a variable q: $a_0 + a_1 q + a_2 q^2 + \cdots$. Not every such infinite series is a modular form. The coefficients a_i have to be formed according to certain rules. The a_i are called the *Fourier coefficients* of the modular form because the series in q can be thought of as the Fourier series of a certain function. (The use of the letter q here is traditional and it should not be confused with a prime number. Therefore, in this chapter we will use ℓ to denote a varying prime number.)

There are particular kinds of modular forms called *cuspidal normalized newforms*. These all have $a_0 = 0$, $a_1 = 1$, and lots of powerful properties.[5] Here is the great fact:

> **FACT**: Every cuspidal normalized newform gives a reciprocity law.

We now explain what this reciprocity law looks like, and then we end with some examples.

---

[4]Notice the word "group" keeps cropping up. Much of what can be done can only be done when there is a group law around.

[5]For example, if m and n share no common prime factor, then $a_{mn} = a_m a_n$.

Because modular forms are related to étale cohomology, you can guess that there will be a whole series of Galois representations in the picture, one for each prime p. Start with a particular prime p.

THEOREM 21.1: If $q + a_2q^2 + a_3q^3 + \cdots$ is a cuspidal normalized newform, then there exist

1. a positive integer N, called the level of the newform;
2. a field k that contains $\mathbf{Q}_p$ and also all the coefficients a_i;
3. a two-dimensional linear Galois representation $r : G \to \mathrm{GL}(2, k)$;

which obey the following rule: If ℓ is any prime that does not evenly divide pN, then r is unramified at ℓ, and $\chi_r(\mathrm{Frob}_\ell) = a_\ell$.[6]

If you are worried about $\mathbf{Q}_p$ and k, we can state a simpler corollary of this theorem, more along the lines of our earlier exposition:

COROLLARY: Given the cuspidal normalized newform as above, assume that all of the a_i's are ordinary integers. Then there exists a two-dimensional linear mod p Galois representation $r : G \to \mathrm{GL}(2, \mathbf{F}_p)$ with the property that

$$\chi_r(\mathrm{Frob}_\ell) \equiv a_\ell \pmod{p}.$$

What should we think about this? The point is that it is not difficult to write down zillions of examples of cuspidal normalized newforms. They, and Theorem 21.1, then immediately tell you that there are zillions of different two-dimensional linear representations of the absolute Galois group G of $\mathbf{Q}$. This means, among other things, that G is very big. And yet you may hope that at least all of its *two*-dimensional linear representations come about like this. That is not true, yet there is an important

[6]Although we know from this theorem that the Galois representations in (3) exist, we cannot write them down explicitly, because we do not have a good understanding of the elements of the absolute Galois group of $\mathbf{Q}$, one by one.

conjecture that has been made which would tell us exactly which two-dimensional Galois representations r to $\mathrm{GL}(2,\mathbf{F}_p)$ come from cuspidal normalized newforms as in the corollary. It is called Serre's Conjecture (after Jean-Pierre Serre, the mathematician who proposed it). The main condition is that the determinant of the 2-by-2 matrix $r(c)$ should equal -1. (Here c once again stands for complex conjugation.) Recent advances have made it plausible to hope that Serre's Conjecture may be proven in the near future.

We give an example. Let

$$\Delta = q(1-q)^{24}(1-q^2)^{24}(1-q^3)^{24}\cdots .$$

Granted, this is not an infinite sum, but instead an infinite product. Still, you can multiply out all those twenty-fourth powers and then multiply out the resulting infinite series in powers of q, and get in the end a single infinite series in powers of q. Although we cannot write out the whole thing, we can compute as many of the coefficients as we like. Traditionally the Fourier coefficients of Δ are called τ_n (Ramanujan's tau function). So $\Delta = q + \tau_2 q^2 + \tau_3 q^3 + \cdots$. The first few values of τ are $\tau_2 = -24$, $\tau_3 = 252$, $\tau_4 = 1{,}472$, $\tau_5 = 4{,}830$, $\tau_6 = -6{,}048$, and $\tau_7 = -16{,}744$.[7] It turns out that Δ is a cuspidal normalized newform whose properties are related to elliptic curves. The level N for this newform is 1 (a rather rare occurrence).

Pick your favorite prime, say 7. Then there is some giant **Z**-polynomial f (which would be very hard to write down) and a representation of its Galois group $r : G(f) \to \mathrm{GL}(2,\mathbf{F}_7)$ with the following property: If ℓ is any prime not equal to 7, then

$$\chi_r(\mathrm{Frob}_\ell) \equiv a_\ell \pmod{7}.$$

For instance, $\chi_r(\mathrm{Frob}_2) \equiv -24 \pmod{7}$.

A pattern like this may not be everyone's cup of tea, but our hope is that if you have read this far in our book, you may be starting to develop a taste for this sort of amazing thing.

---

[7]To compute τ_n for a particular n, use the binomial theorem to expand the powers of $(1-q^k)^{24}$ for all integers $k \le n$, and then multiply together the resulting polynomials, collect terms and look at the coefficient of q^n.

Review of Reciprocity Laws

Here are the examples of reciprocity laws mentioned in this book:

- Quadratic reciprocity.
- Quadratic reciprocity viewed as reciprocity between Frob_ℓ in a Galois representation and the factorization of $X^2 - W$ modulo ℓ.
- Reciprocity between Frob_ℓ in a Galois representation and the factorization of a general irreducible **Z**-polynomial modulo ℓ.
- Reciprocity between Frob_ℓ in a Galois representation and the solution set of an elliptic curve over $\mathbf{F}_\ell$.
- Reciprocity between Frob_ℓ in a certain kind of two-dimensional Galois representation and a modular form.

From this point of view, the big theorem that implied Fermat's Last Theorem, and which was proved in several papers by Wiles and his colleagues, established a reciprocity law: For any elliptic curve E with coefficients in **Q** there is some modular form $q + a_2q^2 + a_3q^3 + \cdots$ with the property that for any prime ℓ (except for a finite number of exceptions) $a_\ell = 1 + \ell - \#E(\mathbf{F}_\ell)$.[8]

We can give a numerical example of this last kind of reciprocity. Let E be the elliptic curve defined by the equation $y^2 + y = x^3 - x^2$. (We could write an equation for E in the usual form as $y^2 = x^3 - 432x + 8{,}208$, but by using the equation $y^2 + y = x^3 - x^2$, we keep the coefficients a lot smaller and avoid technical problems computing a_2 and a_3.) The only "bad" prime for E is 11, and for any prime ℓ other than 11, we can define a_ℓ by the formula

$$a_\ell = 1 + \ell - \#E(\mathbf{F}_\ell).$$

For example, $E(\mathbf{F}_2) = \{\mathcal{O}, (0,0), (0,1), (1,0), (1,1)\}$, and so $a_2 = 1 + 2 - 5 = -2$. Similarly, $E(\mathbf{F}_3) = \{\mathcal{O}, (0,0), (0,2), (1,0), (1,2)\}$, and so

[8]This is a reciprocity law between the modular form and the Galois representations coming from the torsion points on E.

$a_3 = 1 + 3 - 5 = -1$. One more: $E(\mathbf{F}_5) = \{\mathcal{O}, (0, 0), (0, 4), (1, 0), (1, 4)\}$, and so $a_5 = 1 + 5 - 5 = 1$.[9]

We can also take the following infinite product and expand it to as many terms as we like:

$$\begin{aligned} q\prod_{n=1}^{\infty}(1-q^n)^2(1-q^{11n})^2 &= q(1-q)^2(1-q^2)^2(1-q^3)^2(1-q^4)^2\cdots \\ &\qquad (1-q^{11})^2(1-q^{22})^2(1-q^{33})^2(1-q^{44})^2\cdots \\ &= q + b_2q^2 + b_3q^3 + b_4q^4 + b_5q^5 + \cdots \\ &= q - 2q^2 - q^3 + 2q^4 + q^5 + \cdots . \qquad (21.2) \end{aligned}$$

This power series in q is a cuspidal normalized newform of level 11. The amazing thing is that $b_\ell = a_\ell$ for any prime ℓ other than 11! In other words, to compute the number of solutions to $y^2 + y \equiv x^3 - x^2 \pmod{\ell}$, you can multiply the above product up to and including the terms containing q^ℓ, figure out the coefficient of q^ℓ, and use that number to compute $\#E(\mathbf{F}_\ell)$. You can see that this is true for $\ell = 2$, 3, and 5 from our preceding computations.

A vast generalization of this is the conjecture that there are (similar but more complicated) reciprocity laws between *any* Galois representation coming from the étale cohomology of a **Z**-variety and an appropriate generalization of a modular form, called an "automorphic representation." (An automorphic representation is indeed a kind of group representation, but with a totally different source and target from a Galois representation.) This conjecture is part of a grand system of conjectures known as the "Langlands program," after the Canadian mathematician Robert Langlands.

A Physical Analogy

To conclude, we can draw an analogy between the linear representations of the absolute Galois group G of **Q** and old-fashioned relativistic particle physics.

[9] It is just a coincidence that $\#E(\mathbf{F}_\ell) = 5$ for $\ell = 2$, 3, and 5.

In physics, the key group is L, the Lorentz group of spacetime coordinate transformations. All physical laws must be invariant under L. In quantum mechanics, a system is given by a large set of complex vectors H (actually infinite-dimensional), and a state S of the system is described by x_S, a nonzero vector in H (or more accurately, a line in H). If you act on your system by a transformation of coordinates in spacetime, this changes the way you describe the given state of the system, but not the actual physical state itself. So this change of coordinates must take the vector x_S to a new vector that describes the same state S in the new coordinates. The laws of quantum mechanics say that this gives a linear action of L on H. That representation *is* the system S from this point of view. If the representation cannot be decomposed linearly into simpler ones, we call it "irreducible." The irreducible representations correspond to elementary particles. In this analogy, a reciprocity law in physics would be given by a black box that enables you to compute the observables of the system (such as the energy levels) given the representation of L.

Similarly, the absolute Galois group G has many linear representations, and each one reveals or describes a number-theoretical system. The irreducible representations then play the role of particles.

Chapter 22

FERMAT'S LAST THEOREM AND GENERALIZED FERMAT EQUATIONS

Road Map

In this culminating chapter, we gather the fragmentary references to Fermat's Last Theorem from the preceding chapters and show how the ideas of Galois representations and reciprocity laws played a central role in its proof. We also go beyond Fermat's Last Theorem to consider similar problems, called "generalized Fermat equations" and mention that the same ideas can be used to approach them also. As the equations get more complex, however, new ideas, outside the scope of this book, will probably be necessary to solve them.

This chapter is not meant to be a substitute for any of the popular books that attempt to explain the proof of Fermat's Last Theorem. We will focus on the role played in the proof by the ideas in this book. Nor is there room in this chapter to give an adequate historical account of the proof.[1]

A word of warning: This chapter is necessarily more difficult and complex than any other in the book. All the needed ingredients for understanding it, however, have been introduced in the previous chapters. At whatever level of detail you choose to read it, we hope you will get

---

[1]For sketches of Wiles's proof that mention the shoulders on which he stood, see (Hellegouarch, 2002; Singh, 1997; van der Poorten, 1996).

some idea of how the proof of Fermat's Last Theorem uses the "fearless symmetries" of the Galois group that have been our theme.

The Three Pieces of the Proof

In the last chapter, we gave our last look at reciprocity laws per se. Now we will show how these laws can be applied to prove Fermat's Last Theorem (referred to as FLT throughout this chapter) and similar theorems.

Before Andrew Wiles burst upon the scene, there were a number of proposed strategies for proving FLT. Let us begin by explaining the strategy that Wiles ultimately made to work. This strategy is composed of several difficult theorems plus one big conjecture. The proof was completed by Wiles and Taylor–Wiles by proving enough of the big conjecture to make the whole strategy work.[2]

Before we start, we review the statement of FLT:

> If n is an integer greater than 2, there are no solutions to the equation $x^n + y^n = z^n$, where we require x, y, and z to be nonzero integers.

The first step is to note that if we can prove this for the exponents 4 and odd prime numbers, then it is true for any $n > 2$. Why? Given any integer $n > 2$, there are two possibilities:

1. If n is divisible by 4, we can write $n = 4m$, and then the equation $a^n + b^n = c^n$ becomes $(a^m)^4 + (b^m)^4 = (c^m)^4$. If we have shown that there are no solutions with the exponent 4, then this equation has no solutions.
2. If n is not divisible by 4, then it must be divisible by an odd prime p. Write $n = pm$, and then the equation $a^n + b^n = c^n$ becomes $(a^m)^p + (b^m)^p = (c^m)^p$. If we have shown that there are no solutions with the exponent p, then this equation has no solutions.

[2]Wiles collaborated with Richard Taylor to prove one key step in the whole proof. Thus, the proof of FLT appeared in two papers, one by Wiles (Wiles, 1995) and a shorter one that was a joint work of Taylor and Wiles (Taylor and Wiles, 1995).

The equation $a^4 + b^4 = c^4$ was shown to have no solutions by Fermat, using a method called "infinite descent." So, throughout this chapter we will fix an odd prime p. In fact, for technical reasons, we will assume that $p \geq 5$. This is not a problem because FLT had been proven for exponents $n = 3$ as well as $n = 4$ since at least 1770. Because we have fixed the prime p, we have to use other letters for other prime numbers that we will need to use. We will use the letters v, w, and ℓ for these.

There are three basic components to the winning strategy:

1. Frey curves;
2. The Modularity Conjecture;
3. The Level Lowering Theorem.

The first two of these components can be explained without mentioning Galois representations. But explaining the third component requires a reciprocity law (Theorem 21.1). This reciprocity law will then continue to be essential to Wiles's successful execution of the strategy.

Another feature of Wiles's work is that it requires us to enlarge our menagerie of linear Galois representations beyond those that take values in $\mathrm{GL}(2, \mathbf{F}_p)$. It is essential to use Galois representations that take values in matrices with entries from the big field $\mathbf{Q}_p$ (and even more complicated algebraic systems that we will not be able to discuss in any detail). We have already introduced examples of these Galois representations in Theorem 21.1.

From now on, new ideas and concepts are going to come thick and fast. To explain them in detail would take a whole additional book.

Frey Curves

Start with the innocuous looking equation

$$A + B = C$$

where A, B, and C are integers. It is not difficult to solve this equation! But if you put some additional requirements on A, B, and C, it can become difficult. For example, if we require A, B, and C

all to be nonzero integral perfect pth powers, then a solution to $A + B = C$ would be a counterexample to FLT.

Given any solution to $A + B = C$, we can form an elliptic curve, called a "Frey curve" (named after Gerhard Frey)

$$E : y^2 = x(x - A)(x + B).$$

For any prime v, we can consider the two-dimensional Galois representation ψ made with the v-torsion points on $E(\mathbf{C})$ as in chapter 18. Then we can try to find out where ψ is ramified, that is, what are the bad primes for ψ (see chapter 16). It can be proven that any bad prime for ψ either divides the product ABC or equals v or equals 2.

When A, B, and C satisfy certain extra properties, the set of bad primes for ψ can be reduced drastically. For example, if $A = a^p$, $B = b^p$, and $C = c^p$ are all nonzero integral pth powers, and if we assume that $a \equiv 3 \pmod 4$ and b is even, and that a, b, and c have no common prime factor, and if we take $v = p$, it turns out that the only bad primes for ψ are 2 and p!

This is a great result, because in that case, we do not have to know what a, b, and c are in order to know exactly the set of bad primes for ψ. Because we are trying to prove there are no such a, b, and c, it is nice not to need to know what they are, as we try to prove a contradiction from their hypothetical existence.

By the way, it is not hard to prove the following:

LEMMA 22.1: If the Fermat equation $x^p + y^p = z^p$ has a solution with x, y, and z all nonzero integers, then it has such a solution with $x \equiv 3 \pmod 4$ and y even.

If we want to make explicit the dependence of ψ on the parameters A, B, and C, we will denote it by $\psi_{A,B,C}$.

The Modularity Conjecture

In chapter 21, we briefly mentioned modular forms. A *cuspidal normalized newform* is an infinite series[3] $f = q + a_2q^2 + a_3q^3 + \cdots$

[3]This infinite series is called the "q-expansion of f."

that satisfies some complicated symmetry conditions we did not specify and which are beyond the scope of this book. Remember that f comes with a positive integer N called its *level*. We did not mention it in chapter 21, but it also comes with another positive integer k called the *weight* of the modular form. Both N and k appear as parameters in the complicated symmetry conditions we did not specify.

Given two integers N and k, there are only finitely many cuspidal normalized newforms with level N and weight k. Moreover, there is a relatively simple algorithm, suitable for computers, with which you can find them all, as long as N and k are not too big.[4]

So a cuspidal normalized newform gives you a series of integers a_ℓ for all primes ℓ: Just read them off its q-expansion. We call these $a_\ell(f)$ to remind us that they come from the modular form f.

Now remember that an elliptic curve E also gives you a series of integers a_ℓ for all primes ℓ via the equation $\#E(\mathbf{F}_\ell) = 1 + \ell - a_\ell$. We will call these $a_\ell(E)$ to remind us that they come from the elliptic curve E.

Although we never mentioned it, an elliptic curve has an extra important property: It comes with a positive integer N, called its "conductor." There is an algorithm for finding the conductor, given E. There is a theorem that says if ℓ is a prime not dividing υN, then ℓ is a good prime for the Galois representation $G \to \mathrm{GL}(2, \mathbf{F}_\upsilon)$ obtained from the υ-torsion points of $E(\mathbf{C})$. (As usual, G is the absolute Galois group of $\mathbf{Q}$.)

We can now state the Modularity Conjecture.

CONJECTURE 22.2 (*The Modularity Conjecture*): Given any elliptic curve E with conductor N there exists a cuspidal normalized newform f with level N and weight 2 such that the integers $a_\ell(E) = a_\ell(f)$ for all but finitely many primes ℓ.

In this form, the conjecture is rather mysterious. It can be explained in a purely geometric way by using non-Euclidean geometry

---

[4]On the other hand, if you only fix N, there will be infinitely many cuspidal normalized newforms with level N and various weights. If you only fix k, again there will be infinitely many cuspidal normalized newforms with weight k and various levels.

and complex numbers. This approach is discussed in most of the popular books about Wiles's proof of FLT.

For example, the elliptic curve defined by $y^2 + y = x^3 - x$ satisfies the Modularity Conjecture where the corresponding modular form is given by the power series in equation (21.2).

By the way, the Modularity Conjecture is not a conjecture any more. Following the ideas of Wiles and Taylor–Wiles, and with a lot of extra hard work, the proof was completed by Christophe Breuil, Brian Conrad, Fred Diamond, and Richard Taylor (Breuil et al., 2001). However, at the time of Wiles's work on FLT, it was still a conjecture, so we will continue to refer to it as such in this chapter.

Because we cannot go into the details of what a cuspidal normalized newform f really is, the main things that you should keep in mind are:

- f is determined by its q-expansion;[5]
- compared with Galois representations, newforms are easy to compute and work with.

Lowering the Level

Suppose we have a cuspidal normalized newform f of level N and weight 2. By the corollary to Theorem 21.1, for any choice of a prime v, there is a Galois representation $r : G \to \mathrm{GL}(2, \mathbf{F}_v)$ whose traces at Frobenius elements give you the values of coefficients in the q-expansion of f modulo v. We will change the name of the representation from r to ψ_f to emphasize its dependence on f.

To state the Level Lowering Theorem and also to explain Wiles's main contribution, we have to introduce the concept of "irreducibility" of a representation. This is a technical concept and you can skip it on a first reading. But we want to include it so that we can state the theorems in the chapter accurately.

You will need to review the definitions in the first section of chapter 20. Suppose k is a field.

[5]In fact, it is enough to know the integers $a_\ell(f)$ for all primes ℓ to determine the entire q-expansion, and hence determine f.

DEFINITION: A Galois representation $R : G \to \mathrm{GL}(2, k)$ is *irreducible* if there is no line L in the vector space k^2 with the property that $R(\sigma)L = L$ for all σ in G.

Now let k' be a field containing k with the property that any polynomial with coefficients in k' has a root in k'. (For any field k, there always exists such a field k'—in fact, there are many of them.) Note that because every element of k is also in k', we can consider a Galois representation $R : G \to \mathrm{GL}(2, k)$ with values in k-matrices as also defining a Galois representation $R' : G \to \mathrm{GL}(2, k')$ with values in k'-matrices.

DEFINITION: A Galois representation $R : G \to \mathrm{GL}(2, k)$ is *absolutely irreducible* if the corresponding R' is irreducible.[6]

For example, if $R : G \to \mathrm{GL}(2, k)$, and if $R(\sigma)$ takes on all possible values in $\mathrm{GL}(2, k)$ as σ runs through all the elements of the absolute Galois group G, then R is absolutely irreducible. (This is not obvious.)

By the way, for all the elliptic curves E that are used in this chapter, it turns out that if $v \geq 3$, then the representation on the v-torsion points of $E(\mathbf{C})$ is absolutely irreducible if and only if it is irreducible. (So why did we introduce both concepts? For truth in advertising. The "if and only if" we just stated is a difficult theorem.)

Ken Ribet (Ribet, 1990) proved the following:

THEOREM 22.3 (*Level Lowering Theorem*): Let f be a cuspidal normalized newform of level N and weight 2. Suppose that ℓ is a prime dividing N but that ℓ^2 does not divide N, and suppose that $\psi_f : G \to \mathrm{GL}(2, \mathbf{F}_v)$ is absolutely irreducible. Assume further that either

- $\ell \neq v$ and ψ_f is unramified at ℓ (i.e., ℓ is a good prime for ψ_f), or

[6]It can be proven that this concept does not depend on the choice of the field k'.

- $\ell = v$ and ψ_f is "flat" at v.[7]

Then there exists a cuspidal normalized newform g of level N/ℓ and weight 2 with $\psi_g = \psi_f$.

Note that the equation $\psi_g = \psi_f$ tells you about the coefficients in the q-expansions of g and f *after they are reduced* (mod v). In other words, if w is a prime that is not a factor of vN, then $a_w(g) \equiv a_w(f)$ (mod v).

Proof of FLT Given the Truth of the Modularity Conjecture for Certain Elliptic Curves

In this section, set $v = p$, and suppose that FLT is false. Then by Lemma 22.1, there exist nonzero integers a, b, and c such that $a \equiv 3$ (mod 4), b is even, and $a^p + b^p = c^p$. We want to derive a contradiction from this equation, and thereby prove FLT. (Remember we are assuming that p is a prime greater than 3.) Form the Frey curve $E = E_{a^p,b^p,c^p}$ and consider the Galois representation ψ_E obtained from the p-torsion points of $E(\mathbf{C})$. It is known that ψ_E obeys the hypotheses of the Level Lowering Theorem.[8]

The conductor N of E can be computed to be the product of all primes dividing abc. By the Modularity Conjecture, there is a cuspidal normalized newform f of level N and weight 2 such that for all primes w that are not factors of N, $a_w(f) = a_w(E)$, and hence these pairs of integers are also congruent modulo p.

Using a standard theorem of algebraic number theory, it follows from the congruences $a_w(f) \equiv a_w(E)$ (mod p) that the Galois representation ψ_f is equivalent to ψ_E. Using our detailed knowledge of how we constructed the Frey curve E, it can be shown that ψ_E,

---

[7] This is a condition too hard to define in this book, but it more or less means "as well-behaved as possible" if $\ell = v$—it is too much to expect ψ_f to be unramified at v.

[8] In particular, the absolute irreducibility follows from a theorem proved by Barry Mazur in the 1970s. Mazur's theorem had to do with studying the rational solutions to a *different* **Z**-variety, called a "modular curve." Many of his ideas (including the concept of "deformation" discussed later in this chapter) were used in the eventual proof of FLT. Unfortunately, modular curves would require a very long digression to explain.

and hence ψ_f, is unramified at all primes other than 2 and p and flat at p. Using the Level Lowering Theorem, there is a cuspidal normalized newform g of level 2 and weight 2.

The punch line: We compute the set of cuspidal normalized newforms g of level 2 and weight 2, and find out that there aren't any! They do not exist! This is a logical contradiction, which stemmed from our assuming the existence of a, b, and c such that $a^p + b^p = c^p$. Therefore, no such a, b, and c can exist and FLT is proved.

The only weak link in this proof (before Wiles) was that the Modularity Conjecture had not been proven yet. Wiles and Taylor–Wiles proved a large part of the Modularity Conjecture—enough to make this strategy work.

Bring on the Reciprocity Laws

As we have mentioned, to go farther and explain what Wiles did, we have to review the reciprocity laws connected with elliptic curves and modular forms. However, we will have to up the ante and discuss Galois representations not just to $\mathrm{GL}(2, \mathbf{F}_\upsilon)$ but to $\mathrm{GL}(2, \mathbf{Q}_\upsilon)$ where $\mathbf{Q}_\upsilon$ is the field of υ-adic numbers, briefly introduced in chapter 20.

Given an elliptic curve E and a prime υ, in chapter 18 we obtained a two-dimensional Galois representation $\psi_E : G \to \mathrm{GL}(2, \mathbf{F}_\upsilon)$ on the υ-torsion points of $E(\mathbf{C})$. However, using the υ^2-torsion, the υ^3-torsion, and so on, we can get a two-dimensional Galois representation $\Psi_E : G \to \mathrm{GL}(2, \mathbf{Q}_\upsilon)$, where the matrices now have entries in the big field $\mathbf{Q}_\upsilon$. This Galois representation obeys the following reciprocity law: If ℓ is a good prime for Ψ_E, then $\chi_{\Psi_E}(\mathrm{Frob}_\ell) = 1 + \ell - \#E(\mathbf{F}_\ell)$ exactly (not just modulo υ—remember that $\mathbf{Z}$ is a subset of $\mathbf{Q}_\upsilon$ for every υ.)

There is a way to reduce Ψ_E modulo υ. If you write elements in $\mathbf{Q}_\upsilon$ as infinite υ-adic expansions, as in the footnote on page 229, then you can define $\mathbf{Z}_\upsilon$ as the set of all elements in $\mathbf{Q}_\upsilon$ with no digits to the right of the "decimal point." You can reduce an element $\cdots a_3a_2a_1a_0$ of $\mathbf{Z}_\upsilon$ by sending it to a_0 modulo υ. Now it turns out that for any element σ in the absolute Galois group G, $\Psi_E(\sigma)$ is a 2-by-2 matrix with entries in $\mathbf{Z}_v$. So you can reduce each of these entries and get a 2-by-2 matrix with entries in $\mathbf{F}_v$. This defines the reduction of Ψ_E modulo v. We will denote the reduction by $\overline{\Psi}_E$. It is itself a Galois representation $\overline{\Psi}_E : G \to \mathrm{GL}(2, \mathbf{F}_v)$. It turns out that $\overline{\Psi}_E$ is actually equal to ψ_E.

Notice the system behind this notation: Galois representations to $\mathrm{GL}(2, \mathbf{F}_v)$ are denoted by small Greek letters. Galois representations to $\mathrm{GL}(2, \mathbf{Q}_v)$ are denoted by large Greek letters. The reduction of a Galois representation is denoted by putting a bar over the large Greek letter. And if the uppercase and lowercase Greek letters are the same letter, as, for example, Ψ and ψ, we mean to imply that $\overline{\Psi} = \psi$.

On the other hand, if $f = q + a_2q^2 + a_3q^3 + \cdots$ is a cuspidal normalized newform and if all the coefficients a_i are integers, and if v is any prime, then by Theorem 21.1 f has a level N and a two-dimensional Galois representation $G \to \mathrm{GL}(2, \mathbf{Q}_v)$ which we will call Ψ_f. If ℓ is a prime not dividing vN, then the trace of $\Psi_f(\mathrm{Frob}_\ell)$ equals the integer a_ℓ.

As with Ψ_E, we can reduce Ψ_f modulo v to obtain $\overline{\Psi}_f$. It is also a Galois representation $\overline{\Psi}_f : G \to \mathrm{GL}(2, \mathbf{F}_v)$.

We can now state the Modularity Conjecture in a different form that can be proven to be logically equivalent to the original Modularity Conjecture:

CONJECTURE 22.4: Given an elliptic curve E of conductor N, then there exists a cuspidal normalized newform f with level N and weight 2 such that $\Psi_E = \Psi_f$.[9]

If an elliptic curve E satisfies Conjecture 22.4 we say it is modular. Similarly, if a Galois representation $\Phi : G \to \mathrm{GL}(2, \mathbf{Q}_v)$ is equal to Ψ_f for some cuspidal normalized newform f, we say Φ is

[9]Remember the discussion of "equivalent" linear representations from chapter 15? A more accurate way of stating the conclusion of this version of the Modularity Conjecture is to say that Ψ_E and Ψ_f are equivalent. If that is the case, however, then we can make certain choices so that in fact they are equal.

modular. And if a Galois representation $\phi : G \to \mathrm{GL}(2, \mathbf{F}_v)$ is equal to $\overline{\Psi}_f$ for some cuspidal normalized newform f, we say ϕ is modular.

We will have to consider the possibility of using different primes v, so if we want to emphasize which v we are using to define the various Galois representations, we will put them into a subscript on the Greek letter. Then we have the following:

THEOREM 22.5: If $\Psi_{E,v}$ is modular for some prime v, then $\Psi_{E,w}$ is modular for any prime w.

PROOF: If $\Psi_{E,v}$ is modular, then there exists a modular form f such that $\Psi_{E,v} = \Psi_{f,v}$. But it is a general fact (proven in algebraic number theory) that $\Psi_{E,v} = \Psi_{f,v}$ if and only if $a_\ell(f) = a_\ell(E)$ in $\mathbf{Q}_v$ for all but a finite number of primes ℓ. But $a_\ell(f)$ and $a_\ell(E)$ are ordinary integers, so if they are equal in $\mathbf{Q}_v$, then they are also equal in $\mathbf{Q}_w$ for any prime w. Therefore, $\Psi_{E,w} = \Psi_{f,w}$ for any prime w.

What Wiles and Taylor–Wiles Did

Wiles and Taylor–Wiles proved the modularity conjecture for a certain kind of elliptic curve that includes all of the Frey curves. Starting with one of these elliptic curves E and an odd prime v, consider $\Psi_{E,v}$. We want to show that it is modular. So we consider "deformations" of $\psi_{E,v}$, that is, Galois representations $\Phi : G \to \mathrm{GL}(2, \mathbf{Q}_v)$ such that $\overline{\Phi} = \psi_{E,v}$. (The use of the odd term "deformation" for this concept comes from an analogy with a previous use of the term in algebraic geometry.)

We say that a deformation is a *modular deformation* if $\Phi = \Psi_{f,v}$ for some cuspidal normalized newform f of weight 2. Now, it follows immediately from our definitions that $\Psi_{E,v}$ is a deformation of $\psi_{E,v}$. If we can prove that *every* deformation of $\psi_{E,v}$ is modular, we are done, because then $\Psi_{E,v} = \Psi_{f,v}$ for some f, which is what Conjecture 22.4 is asserting. (We have to keep track of the level of f, of course, but we omit mentioning it from now on to streamline the discussion.)

Now let v be any odd prime. There is a way to package all the deformations of a given Galois representation $\psi : G \to \mathrm{GL}(2, \mathbf{F}_v)$ into a single highly complicated Galois representation $R_{\text{univ}} : G \to \mathrm{GL}(2, A_{\text{univ}})$. Here the subscript refers to the word "universal," because this Galois representation controls all the deformations of ψ. The symbol A_{univ} stands for the algebraic object that contains the entries in the matrices $R_{\text{univ}}(\sigma)$ for σ any element in G. This algebraic object is something rather more complicated than a field, called a "complete noetherian local ring" (the definition of which is beyond the scope of this book).

Actually, we lied: R_{univ} does not control *all* the deformations of ψ, but only those satisfying some highly technical conditions. One of the many difficult tasks that Wiles solved was figuring out which conditions to use.

There is also a way to package all the *modular* deformations of a given Galois representation $\psi : G \to \mathrm{GL}(2, \mathbf{F}_v)$ into another single highly complicated Galois representation $R_{\text{mod}} : G \to \mathrm{GL}(2, A_{\text{mod}})$. Here the subscript refers to the word "modular" for obvious reasons. Again, A_{mod} is a complete noetherian local ring.

The main theorem of Wiles and Taylor–Wiles can thus be stated:

THEOREM 22.6: If $\psi : G \to \mathrm{GL}(2, \mathbf{F}_v)$ satisfies certain hypotheses (H), which include being irreducible and being modular, then $A_{\text{univ}} = A_{\text{mod}}$. Hence every deformation of ψ that satisfies the highly technical conditions above is modular.

It follows from Theorem 22.6 that if E is a Frey curve and Ψ is a deformation of $\psi_{E,v}$ satisfying those highly technical conditions, then Ψ is modular. It so happens that $\Psi_{E,v}$ does satisfy those highly technical conditions, so it is modular and therefore the Modularity Conjecture holds for E. And that implies FLT, as we have seen already.

But wait! Not so fast! To apply the main theorem, we need to know that $\psi_{E,v}$ satisfies all of the hypotheses (H). The hypotheses we did not state hold all right, but being irreducible and being modular are not obvious for $\psi_{E,v}$. In fact, $\psi_{E,v}$ can sometimes fail to be irreducible. And its being modular is the mod v version of what we are trying to prove in the first place, so it would seem unlikely that we could verify it a priori.

This is where the miracles start happening.[10] By a theorem of Robert Langlands and Jerry Tunnell (Tunnell, 1981) from the theory of "automorphic representations," we know, by completely different methods from anything discussed in this book that if $\upsilon = 3$, then $\psi_{E,3}$ is automatically modular! So if $\psi_{E,3}$ is irreducible, we can apply Theorem 22.6 for $\upsilon = 3$, and so $\Psi_{E,3}$ is modular.

But what if $\psi_{E,3}$ is not irreducible? Wiles came up with a very clever argument that if $\psi_{E,3}$ is not irreducible, there is *another* elliptic curve E' for which $\psi_{E',3}$ *is* irreducible, and for which $\psi_{E,5} = \psi_{E',5}$. It follows that $\Psi_{E'}$ is modular and hence that $\psi_{E',5}$ is modular and hence that $\psi_{E,5}$ is modular. It also turns out that if $\psi_{E,3}$ is not irreducible, $\psi_{E,5}$ must be irreducible! So now we can apply Theorem 22.6 for $\upsilon = 5$.

Either way, the Frey curve turns out to satisfy the Modularity Conjecture and FLT is proven true. Perhaps even more significant than FLT is the Modularity Conjecture itself, which as we mentioned has now been proven for all elliptic curves.

Generalized Fermat Equations

Now that FLT has been proven, number theorists are moving on to other equations. An equation of the form

$$x^p + y^q = z^r$$

where x, y, and z are unknown integers, and where the three exponents p, q, and r are not all the same, is called a generalized Fermat equation. Because it looks like the equation in FLT, it might be expected that similar methods could be used to solve it. This is partially true. But the fact that the three exponents p, q, and r are not all the same makes it even more difficult than FLT.

To make progress on these generalized Fermat equations, mathematicians usually make some assumptions to narrow the scope of

[10]We mean "start" from a logical, not historical, point of view; these miraculous mathematical facts were known before Wiles proved his theorem.

the problem. For the rest of this chapter we will assume that p, q, and r are all prime and that p is odd.

You can come up with "stupid" solutions to generalized Fermat equations if you allow x, y, and z to share a common prime factor.[11] So from now on, we will assume that x, y, and z share no common prime factor. We also assume that x, y, and z are all nonzero, because solutions where one or more of the variables equals 0 are easy to understand. We call solutions satisfying these assumptions nontrivial primitive solutions. Our general feeling is that, just as for FLT, there should be no nontrivial primitive solutions. We are very far from knowing how to prove this, but some partial results are known, examples of which are described in the rest of this chapter and the next one.

What Henri Darmon and Loïc Merel Did

Darmon and Merel proved that

$$x^p + y^p = z^r$$

has no nontrivial primitive solutions if $r = 2$ and $p \geq 5$ or if $r = 3$ and $p \geq 3$. (Notice that $1^3 + 2^3 = 3^2$ gives a solution for $r = 2$ and $p = 3$.)

Their methods follow the same general pattern as those of Wiles, but various complications arise that limit the exponents to those stated. For example, if (a, b, c) is a nontrivial primitive solution, of $x^p + y^p = z^r$, instead of the Frey curve, they consider the elliptic curves

$$y^2 = x^3 + 2cx^2 + a^p x$$

if $r = 2$ and

$$y^2 = (x + 2c)(c^r(x^3 - 3c^2 x) - 2(a^p - b^p))$$

[11]For example, let a and b be any positive integers, and let $c = a^p + b^p$. Then $x = ac$, $y = bc$, and $z = c^{\frac{p+1}{2}}$ solves $x^p + y^p = z^2$.

if $r = 3$. (This last curve does not look as if it fits our definition of an elliptic curve, but a certain change of variables shows that it is one, in fact.) But if $r > 3$, elliptic curves do not seem to work, and more complicated varieties need to be employed.

Prospects for Solving the Generalized Fermat Equations

Darmon has gone on to sketch a method to solve other generalized Fermat equations but the various conjectures needed to make this method work have not yet been proven. It seems that some new ingredients will be needed. Indeed, other people are carrying on the work with various new ingredients. Here is a quotation from an e-mail message we received from Darmon on February 9, 2005:

> I had the impression . . . that further progress would have to wait for someone more clever, or bringing in additional tools, to come along. Fortunately, people like that did step in. There's been some nice work on generalised Fermat equations by Alain Kraus . . . .
>
> Bennett and Skinner have also done some interesting work, and Ellenberg as well [see chapter 23] . . . .
>
> There has also been very interesting work of Bugeaud, Mignotte, and Siksek combining the modular forms techniques with more traditional approaches like linear forms in logs to solve striking open problems, like the complete list of perfect powers in certain binary recurrence sequences (e.g., the Fibonacci and Lucas sequence). I suspect that this is where the future of the subject lies: Both the methods based on modular forms, and more traditional approaches, run into serious obstacles when dealing with natural Diophantine equations, but because those methods are so different, the obstacles one encounters are likely to be different, so one can hope that the information gleaned from a combination of approaches can be stronger and lead to a solution, where no technique applied by itself could.

Chapter 23

RETROSPECT

Road Map

In this last chapter, we look back and relate everything to our initial goal of solving systems of **Z**-equations. We include a few more explicit examples of systems that can be studied using reciprocity laws with modular forms.

We take note of how this initial goal has shifted somewhat over the course of our journey, and how it has turned into a quest for reciprocity laws.

After a digression on why mathematics is worthwhile, or what motivates people to be interested in mathematics, we end the book with a brief discussion that attempts to look past the frontier of current research into the still murky realm of Galois representations.

Topics Covered

Now that we have been over the sometimes rocky, sometimes (we hope) beguiling road of this book—there being no Royal Road to mathematics—it is time to look back and see where we have been, and also to take a glimpse of the view in front of us. Our goal was to give you some idea of the absolute Galois group of **Q** and its representations. We defined and explained basic mathematical concepts needed before we could even begin our discussion: sets, groups, various number systems, matrices, and representations. These concepts are not confined to number theory. They are useful in many parts of mathematics, science, and technology.

On a parallel track, we introduced concepts specific to number theory: systems of **Z**-equations, the integers modulo a prime number, the Galois group of a polynomial, and the monarch of our whole domain—the absolute Galois group of **Q**. We tried especially hard to give you a feeling for what this group consists of: permutations of the field of algebraic numbers that "preserve" addition and multiplication, and hence permute the algebraic solutions of any given system of **Z**-polynomial equations.

With these basic ideas under our belts, we proceeded to the subject matter proper: the matrix representations of the absolute Galois group of **Q** and how these are used to state (and eventually prove) reciprocity laws. We gave examples stemming from quadratic reciprocity, roots of unity, and torsion points on elliptic curves.

We also tried to explain a bigger picture where, by necessity, we were vague. This involved more advanced concepts, such as $\mathbf{Q}_p$, étale cohomology, and generalized reciprocity laws. We can only hope we were able to give you some intimation of what is involved here, like a physicist trying to explain string theory to the general public. The physicist, however, has the intuitively clear physical analogies of strings and particles to go on, and we have only mountains of abstractions to climb.

Back to Solving Equations

Along the way, we kept telling you that the guiding problem in this branch of number theory is the study of solution sets of systems of **Z**-equations in various number systems. Now we should say a little more about how generalized reciprocity laws help in this project.

Which systems of **Z**-equations would we like to solve? They are likely to be elegant or historical. For example, $x^{13}+y^{13}=z^{13}$ is both. It is an instance of Fermat's equation, the one Fermat's Last Theorem refers to, and it is nicely symmetric and simple. As we have remarked several times in this book, Wiles's proof of Fermat's Last Theorem uses representations of the absolute Galois group of **Q** in a central way.

To be specific, we remember that Fermat's Last Theorem states that if n is an integer greater than 2, then the **Z**-equation $x^n + y^n = z^n$ has no solution where x, y, and z are all integers, except for the trivial solutions where one or more of the variables is set equal to 0.

We can summarize chapter 22 as follows: First, we show that it suffices to assume that n is an odd prime p. Now suppose you have nonzero integers a, b, and c and an odd prime p with $a^p + b^p = c^p$. You can then construct an elliptic curve E (using this hypothetical nontrivial solution of course) in a clever way[1] so that the elliptic curve has the following property: The two-dimensional representation ϕ of the absolute Galois group of **Q** on the ℓ-torsion points of E is such that ϕ should enter into a reciprocity law with a modular form f of level 2 and weight 2. You then compute that a modular form with the properties that f must have cannot exist. The last step—the nonexistence of f—is the easy part. The hard part is establishing the reciprocity law.

Similar equations, of the form $x^r + y^s = z^t$, can also be approached in this way. Besides the examples mentioned in chapter 22, Jordan Ellenberg has shown, by proving and then using some new generalized reciprocity laws, that:

> **THEOREM 23.1**: If x, y, and z are all nonzero and have no factors in common, and p is a prime larger than 211, then there are no integer solutions of the equation
>
> $$x^4 + y^2 = z^p.$$

Here is an even more complicated example of reciprocity: Let S be the **Z**-variety defined by the equation in the five variables X, Y, Z, W, and T:

$$X^5 + Y^5 + Z^5 + W^5 + T^5 - 5(XYZWT) = 0.$$

It has been proven that there is a modular form f whose Fourier coefficients enter into a reciprocity law that enables you to tell, for

[1]In fact, E is given by the equation $y^2 = x(x - a^p)(x + b^p)$.

each prime p not dividing the level, the number of solutions in $S(\mathbf{F}_p)$. For a survey of this and many similar examples, see (Yui, 2003).

Digression: Why Do Math?

We can attempt an answer to the question "Why do math?" as illustrated by the mathematics discussed in this book.

If mathematics is the study of patterns, then we immediately see the two reasons throughout history why people have "done math":

1. Curiosity about patterns, particularly of numbers and shapes, and an æsthetic appreciation of them.
2. The need to study patterns occurring in the "real world."

People sufficiently motivated by the first reason become "pure" mathematicians. They generally cannot understand why everyone does not share their enthusiasm for the beauty of these patterns.

People motivated by the second reason are everyone (including also "pure" mathematicians). Even in mundane affairs, you cannot ignore patterns, such as those in your checkbook or the number patterns in the bus schedule; the arrangement of the rooms and furniture in your apartment (spatial patterns); the knotting of DNA (topological patterns) if you are a molecular biologist, "etc. etc. etc."

We see in these two reasons the sources of "pure" and "applied" mathematics, respectively. This book has concerned itself with a topic in pure mathematics, a subtopic of number theory. Number theory is often thought of as the purest of mathematics. Yet even the ideas in this book, such as prime numbers, étale cohomology, and the corresponding Galois representations, have been finding their applications in the "real world." For example, André Weil's version of the Riemann Hypothesis for curves over $\mathbf{F}_p$ has been applied to the problem of constructing efficient communication networks. The use of prime numbers and their theory in the construction of public key codes is of ever-increasing importance for financial institutions and transactions on the Internet. Number theorists have been starting cryptography companies and earning lots of money—perhaps the clearest indication, in the United States at least, of "reality."

Still, in the end, we find ourselves drawn to the beauty of the patterns themselves, and the amazing fact that we humans are smart enough to prove even a feeble fraction of all possible theorems about them. Often, greater than the contemplation of this beauty for the active mathematician is the excitement of the chase. Trying to discover first what patterns actually do or do not occur, then finding the correct statement of a conjecture, and finally proving it—these things are exhilarating when accomplished successfully. Like all risk-takers, mathematicians labor months or years for these moments of success.

And yet, every encounter with the exotic leads to a lifting of the veil and the destruction of a mystery. Mathematicians often get bored by a problem after they have fully understood it and have given proofs of their conjectures. Sometimes they even forget the precise details of what they have done after the lapse of years, having refocused their interest in another area. The common notion of the mathematician contemplating timeless truths, thinking over the same proof again and again—Euclid looking on beauty bare—is rarely true in any static sense.

Luckily, the mathematical world is so rich and complex that we need not fear running out of patterns to wonder about.

The Congruent Number Problem

For a final example of an interesting problem in number theory that can be attacked using the ideas in this book, we discuss the "congruent number problem." This problem goes back to ancient Greek mathematics, but it was only in 1983 that a lot of progress was made in solving it. It is still not totally solved.

The congruent number problem asks: What integers can be areas of right triangles, all of whose sides are rational numbers? We call these areas "congruent numbers." For example, the 3-4-5 right triangle has area 6, so 6 is a congruent number. The 5-12-13 right triangle has area 30, so 30 is a congruent number. The right triangle with sides $\frac{3}{2}$-$\frac{20}{3}$-$\frac{41}{6}$ has area 5, so 5 is a congruent number. It can be proven that 5 is the smallest congruent number.

The congruent number problem is a generalization of the Diophantine–Pythagorean problem, which we discussed in chapter 9 on elliptic curves. Remember the question: "What are the possible right triangles of area 1 that have all their sides of rational length?" So we were then asking, "Is 1 a congruent number?" Some algebraic manipulations reduced the problem to solving a cubic equation, which is equivalent to studying an elliptic curve. Similarly, for the general congruent number problem, we end up studying certain elliptic curves.

Again, as in the proof of Fermat's Last Theorem, the representations of the absolute Galois group of **Q** that arise from the torsion points on these elliptic curves play a crucial role, as does a certain modular form. Tunnell in (Tunnell, 1983) used these ideas to give an easily computable property that a congruent number n has to satisfy. So we can (provably) rule out a given n, if it fails this criterion. Also, as we mentioned in chapter 20, if a certain conjecture about elliptic curves, called the Birch–Swinnerton-Dyer Conjecture, holds, then Tunnell's criterion is "if and only if" and could be used to rule in or out any given n. Thus, a proof of the Birch–Swinnerton-Dyer Conjecture would also completely settle the congruent number problem.

In both this example and the proof of Fermat's Last Theorem, elliptic curves and the Galois representations that they define are at the forefront. Partly this is because mathematicians understand the relatively simple two-dimensional representations of the absolute Galois group of **Q** that arise in these cases much better than general, higher-dimensional representations. They have proved good reciprocity laws with modular forms for some of these two-dimensional representations, and this has enabled work to proceed to a much deeper level in such problems, than (so far) for general systems of polynomial **Z**-equations.

Peering Past the Frontier

As we move on to other systems of **Z**-equations, we find that particular examples tend to be less and less of interest. Do you really care exactly how many solutions to the system

$$\begin{aligned} 33x^2 - 42xy + z^{72} &= 445 \\ x^2 - 34xyz + z^{333} &= \ \ 12 \\ x \ + y^2 + 2z^3 &= \ \ 11 \end{aligned}$$

there are with x, y, and z all integers? Unless this problem has some burning practical significance, it is not very interesting merely on its own. We just made it up at random. On the other hand, it does represent the kind of problem that interests us in general, namely, finding the solutions to a system of **Z**-equations. Any adequate theory we develop should eventually be able to tell us about this random system and all other systems too.

Thus, our focus tends to shift to whole classes of problems, and therefore to the structural or qualitative statements we can make or prove about them. For example, what computable properties of a system of **Z**-equations can guarantee that there are only finitely many solutions with the variables taking on only rational values? Such questions, which are at the frontier of research, lead back to more structural questions about varieties and the absolute Galois group of **Q**.

For another example, consider the statement: The absolute Galois group of **Q** is big. Now, make it into a precise assertion and prove it. That might include proving that for any positive integer n and any prime p, there is a Galois representation whose image[2] is all of $\mathrm{GL}(n, \mathbf{F}_p)$. (It is not known if this is true.) But there will be many other ways of envisioning the "bigness" of the absolute Galois group of **Q**, according to various other aspects of its structure.

In this way, after we leave behind particular **Z**-equations or systems that have provoked our study in the first place, the structures

[2]The *image* of a group representation ϕ is the set of all elements in the target that are actually values of ϕ.

themselves—the absolute Galois group of **Q**, for example—tend to take center stage. As the structures become more understood, particular problems—the congruent number problem, for example—will achieve their solution. This dialectic between the general and the specific is very common in the history of mathematics.

Thus, the focus of theoretical interest in the study of systems of **Z**-equations tends to shift, as the theory is developed, to Galois groups and their representations. We begin to ask questions: How big can representations of the Absolute Galois group of **Q** be? How can certain families of such representations be parametrized? But if for any reason we need to say something about a particular system of **Z**-equations, the theory should stay closely enough connected to the motivating Diophantine problems to be a powerful tool for attacking these problems.

Bibliography

Artin, Emil. 1998. *Galois Theory*, 2nd ed., Dover, New York. Edited and with a supplemental chapter by Arthur N. Milgram.

Ash, Avner, and Robert Gross. 2000. Generalized non-abelian reciprocity laws: a context for Wiles' proof, *Bull. London Math. Soc.* **32**, no. 4, 385–397.

Ash, Avner, Richard Pinch, and Richard Taylor. 1991. An $\widehat{A_4}$ extension of $\mathbf{Q}$ attached to a nonselfdual automorphic form on GL(3), *Math. Ann.* **291**, no. 4, 753–766.

Ball, W. W. Rouse. 1960. *A Short Account of the History of Mathematics*, Dover, New York.

Boyer, Carl B. 1991. *A History of Mathematics*, 2nd ed., John Wiley, New York. With a foreword by Isaac Asimov. Revised and with a preface by Uta C. Merzbach.

Breuil, Christophe, Brian Conrad, Fred Diamond, and Richard Taylor. 2001. On the modularity of elliptic curves over $\mathbf{Q}$: wild 3-adic exercises, *J. Amer. Math. Soc.* **14**, no. 4, 843–939.

Derbyshire, John. 2003. *Prime Obsession: Bernhard Riemann and the Greatest Unsolved Problem in Mathematics*, Joseph Henry Press, Washington, DC.

Devlin, Keith. 2002. *The Millennium Problems: The Seven Greatest Unsolved Mathematical Puzzles of Our Time*, Basic Books, New York.

Dolci, Danilo. 1959. *Report from Palermo*, Viking, New York. Introduction by Aldous Huxley. Translated from the Italian by P.D. Cummins.

Edwards, Harold M. 1984. *Galois Theory*, Graduate Texts in Mathematics, vol. 101, Springer-Verlag, New York.

Fenrick, Maureen H. 1998. *Introduction to the Galois Correspondence*, 2nd ed., Birkhäuser, Boston, MA.

Gaal, Lisl. 1998. *Classical Galois Theory*, AMS Chelsea Publishing, Providence, RI. Reprint of the third (1979) edition.

Gamow, George. 1989. *One, Two, Three ... Infinity: Facts and Speculations of Science*, Dover, New York.

Garling, D.J.H. 1986. *A Course in Galois Theory*, Cambridge University Press, Cambridge.

Hellegouarch, Yves. 2002. *Invitation to the Mathematics of Fermat–Wiles*, Academic Press, San Diego, CA. Translated from the second (2001) French edition by Leila Schneps.

Hofstadter, Douglas R. 1979. *Gödel, Escher, Bach: An Eternal Golden Braid*, Basic Books, New York.

Klein, Jacob. 1992. *Greek Mathematical Thought and the Origin of Algebra*, Dover, Reprint of the 1968 original.

Koblitz, Neal. 1984. *p-adic Numbers, p-adic Analysis, and Zeta-functions*, 2nd ed., Graduate Texts in Mathematics, vol. 58, Springer-Verlag, New York.

Livio, Mario. 2002. *The Golden Ratio: The Story of Phi, the World's Most Astonishing Number*, Broadway Books, New York.

Mazur, Barry. 2003. *Imagining Numbers (Particularly the Square Root of Minus Fifteen)*, Farrar, Straus, and Giroux, New York.

Nagel, Ernest, and James R. Newman. 2001. *Gödel's Proof*, revised edition, New York University Press, New York. Edited and with a new foreword by Douglas R. Hofstadter.

Nahin, Paul J. 1998. *An Imaginary Tale: The Story of* $\sqrt{-1}$, Princeton University Press, Princeton, NJ.

Penrose, Roger. 2005. *The Road to Reality: A Complete Guide to the Laws of the Universe*, Knopf, New York.

Ribet, K. A. 1990. On modular representations of $\mathrm{Gal}(\overline{\mathbf{Q}}/\mathbf{Q})$ arising from modular forms, *Invent. Math.* **100**, no. 2, 431–476.

Rotman, Joseph. 1998. *Galois Theory*, 2nd ed., Universitext, Springer-Verlag, New York.

Singh, Simon. 1997. *Fermat's Enigma: The Epic Quest to Solve the World's Greatest Mathematical Problem*, Walker and Company, New York. Foreword by John Lynch.

Smith, David Eugene. 1958. *History of Mathematics*, Dover, New York.

Smullyan, Raymond M. 1992. *Gödel's Incompleteness Theorems*, Oxford Logic Guides, vol. 19, Oxford University Press, New York.

Stewart, Ian. 1989. *Galois Theory*, 2nd ed., Chapman and Hall, London.

Struik, Dirk J. 1987. *A Concise History of Mathematics*, 4th ed., Dover, New York.

Taylor, Richard, and Andrew Wiles. 1995. Ring-theoretic properties of certain Hecke algebras, *Ann. of Math.* (2) **141**, no. 3, 553–572.

Tunnell, Jerrold. 1981. Artin's conjecture for representations of octahedral type, *Bull. Amer. Math. Soc. (N.S.)* **5**, no. 2, 173–175.

Tunnell, Jerrold. 1983. A classical Diophantine problem and modular forms of weight 3/2, *Inventiones Math.* **72**, no. 2, 323–334.

van der Poorten, Alf. 1996. *Notes on Fermat's Last Theorem*, Canadian Mathematical Society Series of Monographs and Advanced Texts, Wiley-Interscience, New York.

Weyl, Hermann. 1989. *Symmetry*, Princeton Science Library, Princeton University Press, Princeton, NJ. Reprint of the 1952 original.

Wiles, Andrew. 1995. Modular elliptic curves and Fermat's last theorem, *Ann. of Math.* (2) **141**, no. 3, 443–551.

Yui, Noriko. Update on the modularity of Calabi–Yau varieties, *Calabi–Yau varieties and mirror symmetry* (Toronto, ON, 2001), Fields Inst. Commun., vol. 38, American Mathematical Society, Providence, RI, 2003, pp. 307–362. With an appendix by Helena Verrill.

Index